D1444402

Genetic Control
of Insect Pests

Genetic Control of Insect Pests

G. DAVIDSON

Reader in Entomology as Applied to Malaria,
Ross Institute of Tropical Hygiene,
London School of Hygiene and Tropical Medicine,
London, England

1974

Academic Press

London · New York

A Subsidiary of Harcourt Brace Jovanovich, Publishers

SB 933.3
D38

ACADEMIC PRESS INC. (LONDON) LTD.
24/28 Oval Road,
London NW1

United States Edition published by
ACADEMIC PRESS INC.
111 Fifth Avenue
New York, New York 10003

Copyright © 1974 by
ACADEMIC PRESS INC. (LONDON) LTD.

All Rights Reserved

No part of this book may be reproduced in any form by photostat,
microfilm, or any other means without written permission from the publishers.

Library of Congress Catalog Card Number: 74 5669
ISBN: 0-12-205750-3

PRINTED IN GREAT BRITAIN BY
BILLING & SONS LIMITED, GUILDFORD AND LONDON

PREFACE

The idea of employing genetically manipulated insects to control an insect population evolved from problems created by the use of insecticides. Indeed, resistance to insecticides could be considered as one of the prime reasons for the development of genetic control. An additional impetus was provided by a concern about the growing pollution of the environment caused by the use of persistent toxic chemicals.

The deliberate rearing of sterile insect pests on a large scale and their subsequent release into natural populations, has the added attraction of controlling only those insects which pose a threat. Thus, harmless or beneficial animals of the same ecosystem are not adversely affected, as is often the case with the use of insecticides. Moreover, insect resistance to such control methods seems only a remote possibility.

The initial success of the screw-worm campaign in the United States of America stimulated an interest in the application of such low cost methods to all major agricultural, medical and veterinary pests. This book records the results of laboratory and field trials of different control methods used against some of these pests.

The subject is now passing through a difficult period, in which the general applicability of genetic control is being questioned. Such methods may not succeed when used to control insects of a high biotic potential and which are capable of enormous increases in numbers from very low densities. In addition the logistics of control over large continental areas seem to some to be beyond man's ingenuity. What is needed at this stage in the development of these new methods are carefully planned trials in island situations free from external invasion. Increasing success under these conditions will inevitably lead to more serious consideration for their extension to continental areas.

Having spent a lifetime in the field of tropical medical entomology, the author is still convinced that insecticides have an important role in the control of the insect-borne diseases of man and that, in the way in which they are normally used for this purpose, they contribute little to any pollution of the environment. Of more concern is their cost and the organization required for their efficient application. The ever-increasing threat of resistance and the lack so far of suitable chemical and biological alternatives add urgency to a situation where even now a large section of the world's population is exposed to the ravages of preventable disease. The final solution to pest control must surely come from a combination of methods, strategically applied. One such category is likely to be genetic.

May 1974   G. Davidson

CONTENTS

Preface v

Acknowledgements ix

1. **Introduction** 1

2. **The principles and dynamics involved in the sterile insect technique** 5
    - I   Mass-rearing 6
    - II   Sterilization 14
    - III   Release 17
    - IV   Dynamics 22

3. **Sterilization by irradiation** 31
    - I   Livestock pests 32
    - II   Agricultural pests 38
    - III   Public health pests 45
    - IV   Radiation resistance 47

4. **Chemosterilants** 49
    - I   Public health pests 52
    - II   Agricultural pests 65

5. **Hybrid sterility** 69
    - I   Anopheline mosquitoes 70
    - II   The *Aedes mariae* complex 97
    - III   Tsetse fly crosses 97
    - IV   *Teleogryllus* crosses 98
    - V   Reduviid bug crosses 98

6. **Cytoplasmic incompatibility** 99
    - I   The *Culex pipiens* complex 99
    - II   The *Aedes scutellaris* complex 105

7. **Translocations** 107
    - I   General considerations 107
    - II   Dynamics 113
    - III   The isolation of translocations 116
    - IV   Compound chromosomes 125

vii

8. **Other methods of genetic control** 127
    I   Lethal factors 127
    II Meiotic drive and sex distortion 128
    III Species replacement 130

9. **Summary and conclusion** 133

References...137

Subject Index..149

ACKNOWLEDGEMENTS

I am indebted in the first place to the Dean and to Professor L. J. Bruce-Chwatt of the London School of Hygiene and Tropical Medicine for relieving me of my normal duties while I wrote this book. I am also most grateful to the World Health Organization for their generous permission to quote from mimeographed documents and from the monthly reports of the WHO/ICMR Research Unit on Genetic Control of Mosquitoes and to reproduce Figure 9.

Dr. C. B. Cuellar has been a constant source of encouragement and has been kind enough to allow me to use unpublished information, as also have Dr. R. J. Wood and Miss C. Cubbin. The journal *Nature* was also good enough to allow the reproduction of Figures 2 and 3. Additionally I would like to thank Mr. R. H. Hunt, of De Beers Research Laboratories, Chiredzi, Rhodesia, for permission to reproduce Fig. 5 and Miss J. Chalkley, of the Ross Institute, London, for permission to publish Figure 4.

A book is far from finished when the author has ceased to "scribble". For the final preparation of the manuscript I am most grateful to my colleague Dr. Joan Bryan, aided by the Ross Institute insectary staff, and for the diagrams and photographs to Mr. C. J. Webb and his assistants of the Visual Aids Department at the London School of Hygiene and Tropical Medicine.

1. Introduction

Most people agree that the control of the insect pests of man, his domestic animals and his food crops is a necessity. The control of those insects carrying human disease has led to an alleviation of the sufferings of millions. It has also undoubtedly contributed to an increase in human life expectancy and to the growth in population numbers about which there is so much concern nowadays. While there is considerable argument about whether or not man should impose his own curb on population growth by some form of family planning no-one advocates the abandonment of the control of insect-borne diseases. Many people are of the opinion, in fact, that there is room for many more individuals on this planet and that adequate food resources can be created. It is not the purpose of this book to elaborate this discussion. What seems obvious is that control of the insect pests of food crops and domestic animals is as essential as it ever was if everyone on earth is to have sufficient food. This utopian situation has never been achieved in the whole history of the human race, even in times of much smaller populations.

Insecticides, particularly the stable and persistent ones, have provided a rapid and efficient means of combatting the ravages of all three kinds of insect pests but it is precisely because of their stability and persistence, that problems of resistance and environmental pollution have presented themselves. Insecticide resistance is nature's answer to the widespread use of long-lasting chemicals whose presence constitutes a selective influence killing off those individuals not genetically endowed with protective genes and allowing those that have them to survive and pass them to their offspring. Persistence and wide dissemination combine to maximize this selective influence.

According to Brown (1971) and Brown and Pal (1971) some 130 species of arthropods of agricultural and veterinary importance have shown resistance to insecticides and 102 species of medical importance. Most of these are insects. There are in fact very few of the major insect pests that are not resistant to one insecticide or another in some part of their distribution. Once resistance appears and the insecticide can no longer be used it has been customary to change to another, and with many insects these changes have been so frequent that alternatives are becoming exhausted. The common housefly (*Musca domestica*) is a prime example. Georghiou (1971) cites a single population in California resistant to DDT and to the following organophosphates: malathion, diazinon, ronnel, fenthion, naled, dimethoate, zytron and dichlorvos. In other populations resistance to cyclodiene chlorinated hydrocarbons, e.g., BHC and dieldrin, and also carbamates, e.g., carbaryl, isolan and propoxur is known and there was even a case of resistance to pyrethrins in Sweden (Davies *et al.*, 1958). Among agricultural pests the Egyptian cotton leafworm (*Spodoptera littoralis*) is beginning to compete with the housefly in showing resistance to DDT, cyclodienes, organophosphates and the carbamate, carbaryl.

The use of insecticides in the field of public health has been largely confined to the restricted habitat of the human dwelling and has had comparatively little impact on non-target organisms. This is in complete contrast to their use in agriculture where harmful and beneficial creatures alike have been affected. In fact many of the cases of resistance in insects of medical importance have resulted from the contamination of their breeding places with insecticides used to spray crops. The most recent example is the first recorded instance of mutliple resistance involving organochlorine, organophosphate and carbamate insecticides in the malaria vector, *Anopheles albimanus*. Georghiou (1972) describes a population of this mosquito from El Salvador resistant to DDT, dieldrin, the organophosphates: parathion, methyl parathion, malathion and fenitrothion and the carbamates: propoxur and carbaryl, and attributes the organophosphate and carbamate resistance to the intensive use of these insecticides for the spraying of cotton crops and to a lesser extent of rice and corn. Cotton is treated up to 30 times during the 6-month growing season with such insecticides as parathion, methyl parathion, malathion azodrin, trichlorfon, azinphosmethyl, DDT, carbaryl and others against cotton leafworm, fall armyworm, boll weevil, cabbage looper, *Aphis gossypii* and other pests, and most of the insecticide is applied from the air.

Insecticides, then, have contributed to environmental pollution, though more from their use in agriculture than in the control of insect-borne diseases. Rachel Carson (1962) in writing her book "Silent Spring" brought the seriousness of the situation to the general public in the exaggerated and emotional way of presenting the spectre of a world without birds, bees and butterflies. Appearing at a time when resistance was becoming commonplace it served to reinforce the pressing need for alternative methods to those concerned in the control of insects. However, the readiness with which those concerned took to the use of insecticides after the Second World War was in itself an indication of the relative inefficiency of existing methods. While control operators will undoubtedly take another look at existing and tried methods they will undoubtedly prefer new techniques of comparable efficiency with insecticides.

Special crop culture practices and the search for resistant crop plants will intensify in importance as will general environmental sanitation measures aimed at the reduction of the breeding places of insects of public health importance. These are long-term and often very expensive solutions and far from universally applicable. Biological control methods involving the release of predators, parasites and pathogens have had their successes but there have been many failures also. Knipling(1972) attributes some of these failures to deficiencies in release levels and thinks there is increased likelihood of success now that we know so much more about the mass-rearing of different organisms. Their use is particularly appealing to him as they are methods which are most efficient when pest density is high. However, there is always the uncertainty of whether or not these agents will restrict themselves to the target pests, and in the case of pathogens, and perhaps of parasites too, evidence of a simple genetic basis of susceptibility is accumulating. This implies that refractory individuals already exist and these will of course survive just as insecticide-resistant individuals have done.

Attractant traps have long been recognized as potential safe methods of insect control but really efficient and specific ones are still a rarity. The practical role of natural juvenile hormones and their synthetic analogues is as yet uncertain, and already there are indications of "cross-resistance" to them being shown by insecticide-resistant populations of the housefly (Cerf and Georghiou, 1972) and the flour beetle (*Tribolium castaneum*) (Dyte, 1972).

This book deals with a completely new concept in insect control—the use of insects to control themselves. For the most part it entails the mass-rearing, sterilization and release of populations in the hope that these will mate with wild populations leading to reductions in fertility and perhaps to population elimination. More involved techniques exist which make use of naturally existing population incompatibilities and techniques which can result in population replacement rather than eradication, the intention being to render such replacement populations harmless beforehand by genetic manipulation.

By their very nature these genetic control methods are species specific and non-polluting. Where they lead to population elimination there may be an "upset in the balance of nature" though so far as is known the successful eradication of insect pests in the past by other means has not led to any major catastrophe in this direction. Genetic control methods have the advantage over most other methods of being most efficient when the target insect is in low density as the released insects have the capacity, if they are competitive, to search out the wild populations. However, they are least efficient against those insects with a high reproductive potential. Against such populations they are best used in seasons of low population numbers or in combination with other methods designed to reduce population numbers.

2. The Principles and Dynamics Involved in the Sterile Insect Technique

The genetic control methods to be considered in this book mainly concern the release of sterilized, sterile or incompatible insects into wild populations with the aim of producing a proportion of sterile matings high enough to result in a significant reduction in wild population size and possibly even in its elimination. Exposure to ionizing radiation or to certain chemicals are ways of deliberately sterilizing insects. Hybridizing closely related species is another way of producing sterile insects, while the incompatibility method involves the release of one sex of certain species which is perfectly fertile with its own opposite sex but incompatible with the opposite sex of another population of the same species. While all three methods owe their sterilizing effects to different genetic mechanisms their dynamic effects are identical. As we shall see this is not the case with the other methods to be considered. Translocations, for example, produce an inherited incomplete sterility. Here the dynamics are more complicated and while in theory they can result in population eradication, they seem more likely to result in population replacement, though this in itself may be a useful attribute. In fact deliberate population replacement without the use of translocations will also be considered. Finally, meiotic drive and sex-distortion mechanisms have a peculiar place of their own with some characters in common with all the other methods. Thus what is to be said in this section on principles and dymanics applies in the first instance to deliberate sterilization, hybrid sterility and cytoplasmic incompatibility but will also have considerable relevance to all the other methods.

The principal requirement for success of all genetic control methods must be the production of sufficient numbers of healthy, competitive (though genetically different) insects and their release in the right place at the right time for them to mate successfully with wild insects. Success will depend on a knowledge of how to rear the species on a large scale, how to sterilize or otherwise genetically manipulate without affecting mating ability and competitiveness, and a detailed acquaintance with the general ecology and bionomics of the insect to be controlled.

I. Mass-Rearing

The principle of the sterile insect technique involves the inundation of the wild population with sterile insects. Thus a method of rearing the insect in question in large numbers is an essential. Obviously the ideal insect for such large-scale culture is one with a high biotic potential, with a short life cycle and with simple food requirements. In this context high biotic potential is taken to mean that the female is readily fertilized in captivity, lays large numbers of eggs at frequent intervals and lives for a long time and that the immature stages do not suffer high mortalities. Although it is conceivable that mass-rearing could be accomplished with some species from the continuous capture of wild, fertilized females and the rearing of their progeny, advantages accrue from the setting up of self-perpetuating colonies. One of these is the ability to control the incidence of pathogenic organisms and another is to prevent the occurrence of diapause, common in Lepidoptera for example. Diapause is usually initiated by a change in day-length. Keeping a colony under standard conditions of photoperiod may prevent its occurrence. Colonies can be maintained under "aseptic" conditions, though these can seldom be stringent. Containers and food can be autoclaved and various antimicrobial agents such as antibiotics, fungicides, benzoates, sorbic acid, formalin, mercuric chloride, ethanol, sodium hypochlorate, etc., used either as food additives or for the washing of eggs.

Ideally the food for mass cultures should be nutritionally adequate for the species concerned and attractive to it. It should also be cheap and easily prepared. Different diets may be required for adults and immature stages. Attractants may help to encourage feeding on unfamiliar diets and incitants may be needed to release the biting response. These may be particularly necessary in the case of caterpillars accustomed to feeding on the edge of leaves for example. Feeding stimulants are usually provided by some of the diet constituents. Blood-feeding insects present special though not insurmountable problems. Seldom are they restricted to their usual host however, and membrane feeding on citrated, oxylated, heparinized or defibrinated blood can often be substituted for the use of live animals.

A common difficulty in the establishment and maintenance of colonies is to produce confined conditions allowing natural mating behaviour. Encouragement of mating may be achieved by light quantity and quality adjustments, but as we will see later, selective processes may also be involved which eventually produce alterations in mating behaviour. Oviposition may also present a problem whose solution may lie in experimentation with different oviposition sites or in the use of specific attractants.

For the production of really large numbers, standardization of rearing methods and some system of automation may be necessary. Counting devices ensuring equal densities of individuals to be reared and precise periodic deliveries of quantities of food related to stage of development are some of the requirements, as well as means of separation of the stage needed for release. Where one sex only is to be released accurate mechanical sex separation aids are an advantage.

There can be no doubt that the continuous maintenance of such colonies under such artificial conditions and the attempted production of as a high a yield as possible must lead to the eventual emergence of insects differing in many respects from those occurring under natural conditions. Numerous geneticists have been highly critical of this emphasis on quantity and uniformity e.g., Coluzzi (1971), Boller (1972), Mackauer (1972) and Boesiger (1972). They all emphasize the loss of heterozygosity through colonization. Mackauer (1972) points out the importance of using a large number of individuals from the centre of distribution of the species for the foundation of the colony. Boller (1972) instances the genetical bottleneck through which most colonies pass during their establishment—a difficult initial period followed by adaptation to the artificial conditions, during the course of which there must be considerable loss of genetic diversity. Boesiger (1972) is of the opinion that while initial marked reductions of population size may result from releases of mass-reared sterile insects, difficulties will arise when the population is reduced and dispersed. The remaining insects may be those which have evaded the attentions of the released individuals and he envisages a kind of resistance to genetic control —a selection in the wild for individuals which will not recognize and mate with the introduced ones. Coluzzi (1971), considering the mass-rearing of mosquitoes, favours a diverse founder colony and insectary conditions simulating field conditions as closely as possible at least with regard to daily changes in light and temperature. All these authors comment on the trend away from diversity resulting from inbreeding in colonies. Craig (1964) considering the mass-rearing of *Aedes aegypti* suggests the keeping of two colonies and the crossing of them to produce material for release. This would produce a high degree of hybrid vigour if the two parent colonies have been inbred for some time. As we shall see when we come to consider cytoplasmic incompatibility and translocations, it is possible by outcrossing to wild populations to transfer the genetic modifications involved to a wild genetic background and thus, theoretically at any rate, eliminate many of the differences accumulated during the isolation of the genetic changes.

What has not been considered seriously is the deliberate selection in mass-rearing procedures for those characteristics which might contribute most to the efficiency of the sterile insect technique. The most obvious characteristic is mating efficiency. Apparently it is not difficult to change the mating speed in *Drosophila melanogaster*. Manning (1961) selected both slow and fast lines through 25 generations and produced a difference of 80 min. in the slow line and only three minutes in the fast one. Both sexes were affected by this selection. Personal attempts by the writer to improve mating efficiency in *Anopheles gambiae* species A by selecting females seemed to make little difference in four selections even though the final selection was from a female which was mated within 17 h of emergence from the pupa. It remains to attempt male selection.

Gast (1968) considers mass-rearing from an economic point of view. To him the simple object of the exercise is to produce an acceptable insect at the lowest possible cost. He shows very convincingly that it is cheaper to rear one million boll weevils (*Anthonomus grandis*) using a technique giving only 10% yield than one giving an 80% yield because while egg collection is cheap (10c/1 000) larval diet is expensive—one million from a regime giving a 10% yield cost $1 000 as compared with double that figure for an 80% yield method. Changing larval diets can lead to enormous savings in costs. Substituting cotton seed meal and sugar-cane bagasse for dehydrated carrots and yeast, reduced the cost of rearing the Mediterranean fruit fly (*Ceratitis capitata*) from $80 per million to $4 in one scheme. Colossal savings were made in the same way in the screw-worm campaign as we shall see.

There then are two extreme views—the one concentrating on quantity and costs and the other adhering strictly to the production of insects as similar as possible to the natural ones. In the writer's opinion no rearing method is free from biological criticism and the end product must differ from the wild creature no matter how much care is taken. Lowered competitiveness resulting from colonization and mass-rearing may be compensated for by the weight of numbers of introduced insects. As we shall discuss later the success of such releases may not depend entirely on successful matings between released and natural insect. Overcrowding and aggression effects may contribute. Mass-rearing followed by pilot-release trials designed to assess competitiveness is really the only practical solution. Trying to simulate all the natural conditions could prove more costly than existing methods of control and could result in the release too few insects to have any effect.

Only a few examples of mass-rearing methods will be given here. For further details of insects of medical, veterinary and agricultural importance the reader is referred to a book by Smith (1966) and to various publications by the International Atomic Energy Agency, Vienna (see, References).

The mass-production of the screw-worm (*Cochliomyia hominivorax*) is described in detail by Smith (1967). Adult flies are kept in complete darkness in colony cages of some 50 000–60 000 individuals and the food provided for them is a mixture of ground lean horse meat and honey. To produce 150 million flies per week it is necessary to set up 18 such cages every day. When 8 days old the flies are offered a special oviposition medium and light is admitted. The oviposition medium consists of ground horse meat to which has been added an oviposition stimulant (either extract of heart or of blood albumen). This is presented in a tray containing a heating coil maintaining the temperature around body heat (37–39°C) and a battery of low-powered electric light bulbs (7·5 W) is placed on top of the medium. This acts as a further attractant and oviposition is completed in about 4 hours. Larval rearing is in trays of lean ground meat, blood, water and formaldehyde. In the main rearing plant now at Mission, Texas, the potential exists for the production of 150 million flies a week and for this some 20 000 lb. of dried blood and 200 000 lb. of meat are required. At 4–6 days the larvae are mature and migrate to the sides of the trays and eventually fall into channels of slow moving water which carry them to central collecting points where they are drained off and placed in sawdust to pupate. The pupae are then collected by a sifting process and held for $5\frac{1}{2}$ days at 27°C before irradiation. Gast (1968) refers to a cost of production using this diet of $1 345 per million, and a change to a mixture of blood, fish meal and milk solids leading to a reduction to $800 per million, a saving over a year's production of 150 million flies per week of $330 000.

The fruit flies have proved particularly easy to mass-rear. Nadel and Peleg (1968) describe in some detail the methods used in Israel to rear up to two million Mediterranean fruit flies (*C. capitata*) per day. They used a cage measuring 210 × 30 × 50 cm containing 25 000–50 000 flies as a production unit. Adult food was provided in the form of a mixture of yeast and sugar and oviposition was achieved through the provision of loosely woven cloth through which the female obligingly inserts her ovipositor to let her eggs fall into a collection tray underneath. From three such cages 350 ml of eggs were collected over a period of about two weeks. At 20 000 per ml this represents some 7 million eggs. The larvae were reared in trays measuring 25 × 35 × 1·5 cm containing 1·75 kg of food, at 30 000 per tray. The yield was about 67%. The larval diet was a moist mixture of wheat bran, brewers' yeast and sucrose with two antimicrobial benzoates (Nipagin and Nipasol) added. The pH of the water used was adjusted to 4·3–4·5 by the addition of hydrochloric acid. 10% of the yield was returned to stock and one man could manage the production of nearly two million flies per day at a cost of $5 to $10 per million.

Mosquitoes are also easy to rear on a large scale. This is particularly the case with *Aedes aegypti*, and Smith (1967) describes the methods used to produce more than 10 million males of this species over a period of 10 months at Savannah, Georgia. However, as numerous references will be made in this book to attempts to control *Culex pipiens fatigans* by the Research Unit on the Genetic Control of Mosquitoes at Delhi it is intended to expand on the mass-rearing of this species. This research unit was established by the World Health Organization, in collaboration with the Indian Council of Medical Research, in 1969 to determine the operational feasibility of genetic control techniques for the control or eradication of *C. p. fatigans* and to obtain data on the reproductive biology and population dynamics of *Ae. aegypti* required for the genetic control of this species. The studies are to last over a period of 6–7 years and may culminate in a large-scale attempt to control *C. p. fatigans* in a suburb of the city of Delhi. The anticipated cost of the project is 2·6 million dollars and financial assistance is being given to it by the United States Public Health Service (World Health Organization, 1971).

The unit has a capability of producing some 5 million mosquitoes a week, or some 360 000 male pupae per day.[1] The basic units are cages 72 × 60 × 60 cm stocked initially with 15 000 females and 5 000 males with additions of 3 000 females and 1 000 males on alternate days. 8 such cages are needed. They are kept in a room maintained at 30°C and 80–85% relative humidity with a 13 h light and 11 h dark cycle. 3–4-month-old chickens held still in a restraining device are left on top of the cages overnight on alternate nights to provide blood for the females. On the nights without the chicken ovipositing trays of water which has been previously used for rearing larvae are put into the cage. A total of some 6 000 egg-rafts are obtained each day. The larvae are reared at 31°C in plastic trays 68 × 63 × 9 cm containing 24 l of water at a density of 30 000 per tray. Measurement of numbers involves the use of tubes calibrated from the number of rafts completely filling the surface area of the contained water and assuming 200 eggs per raft. The food consists of equal parts of ground dog biscuit and brewers' yeast and is distributed at 5, 6, 13, 20, 25 and 20 gm/per tray on days 0–5. 204 such trays are available. Pupation occurs on the sixth day and separation from larvae is achieved by straining and immersing in iced water. In such water the larvae sink and the pupae float (Ramakrishna *et al.*, 1963; Weathersby, 1963). As in this species males pupate first and the primary interest is in males for release, pupae are only normally collected on the sixth and seventh days after seeding the larval rearing trays. Collected pupae are then released beneath a grid, the holes in which are of such a size that most of the male pupae pass through but very few of the female ones, and a combination of

[1] Singh, K. R. P., Patterson, R. S., LaBrecque, G. C. and Razdan, R. K. (1972). World Health Organization mimeographed document WHO/VBC/72.386.

2. THE STERILE INSECT TECHNIQUE

this technique and the cropping of pupae on the first two days only produces pupae which are 95–98% male (Sharma *et al.*, 1972). Costs are reckoned at $10 per million exclusive of manpower ($40 inclusive). ·The unit is now organizing for the production of 70 million mosquitoes per week.

Anopheline mosquitoes are somewhat more difficult to mass-rear than *Ae. aegypti* or *C. p. fatigans* but not markedly so. The first species to be released on a large scale, *An. quadrimaculatus*, is now considered to be one of the more difficult. Yet more than 10 years ago nearly half a million males of this species were irradiated and released over a 14 month period (Weidhaas *et al.*, 1962). Since that time *An. stephensi* has been reared on a large scale for chemotherapeutic work (Gerberg *et al.*, 1968) and a very recent field release of chemosterilized *An. albimanus* has taken place in El Salvador involving the production of 100 000–120 000 pupae per day.[1] Though the difference in size of male and female pupae is not as marked in anophelines as it is in culicines it proved possible in this trial to separate fractions containing 86% males using a pupal separator (Fay and Morlan, 1959). Altogether 4 360 000 sterile males were released.

A much more modest production was achieved in the hybrid sterility field trial of Davidson *et al.* (1970). Here two species of the *An. gambiae* complex had to be reared and crossed, in one direction, to produce the sterile males. Cages of 30 × 30 × 30 cm were stocked with 600 pupae and emergent females fed on immobilized rabbits. Oviposition was on to free water containing 7 g/l of sea salt (equivalent to 20% sea water) in the case of *An. melas* and on to filter paper moistened with plain water in the case of *An. gambiae* species B. Larval rearing in flat enamel trays at a density of about one larva per square inch (6·25 cm²) of water surface area, anophelines being mainly surface feeders. *An. melas* had to be reared in 20% sea water; species B in local river water. Food was provided in the form of a finely ground proprietary cereal baby food containing added vitamins and minerals (Farex). Pupation usually started on the eighth day after hatching of the eggs. The pupae were handpicked for the most part. The capacity of the insectary was such that approximately 3 000 *An. melas* and 2 600 species A pupae could be produced each day. Of the 3 000 *An. melas* pupae 600 were used for colony maintenance while the other 2 400 were set aside for emergence and for the isolation of approximately 1 200 virgin females (separated within about 12 h of emergence) for the cross. Of the 2 600 species B pupae, 200 were used for colony maintenance while the other 2 400 were used as the source of approximately 1 200 males for the cross. The cross between species B males and *An. melas* females was then made in cages of 300 males and 300 females. Thus

---

[1] Vector Genetics Information Service, 1972. World Health Organization mimeographed document VBC/G/73.1

at any one time seven cages of *An. melas*, 3 cages of species B and 28 cages of crosses were in existence producing on average something like 6 000 *An. melas* 4 000 species B and 40 000 hybrid eggs per day. Approximately 300 trays with an average surface area of 3 125 cm² were needed to rear these aquatic stages. As on average less than 50% of the hybrid eggs hatched (those destined to become sterile males; only very few females normally resulted from this cross) a maximum yield of hybrid pupae of approximately 15 000 per day was expected, allowing for some aquatic stage mortality. In fact the maximum achieved was 12 000 but the average was only about half this figure. Though these pupae were hand picked they were not counted. An indirect measurement of quantity was achieved using a glass funnel 16 cm in diameter graduated by pouring in lots of 100 pupae in an excess of water and running off the water until the pupae formed a continuous uninterrupted film when at rest. If no shadow was cast and the water was allowed to run out very slowly an indication of the end-point was obtained when surface overcrowding caused sudden violent activity among the pupae. The accuracy of this method was proved by the later counting of pupal skins and dead pupae remaining in sample release containers. 56 651 individuals were counted from these samples. The estimated number from funnel measurements was 54 450. Altogether some 300 000 sterile hybrid males were produced over a period of the two months.

The mass-production of the house fly *Musca domestica* is well illustrated by the description given by Smith (1967) of the field release experiments carried out by Weidhaas and LaBrecque (1970) on Grand Turk Island in the Caribbean. Over a 10-week period 33 million flies were released. Stock cages 50 × 50 × 50 cm contained 20 000 flies fed on a mixture of sugar, dried milk and dried egg yolk. 5 000 new pupae were added twice a week. Oviposition cups were provided containing aged larval food wrapped in moist black cloth. Eggs were washed off this cloth into a measuring cylinder, allowed to settle and 5 ml quantities estimated to contain 30 000 eggs. Such a quantity was put into each of the larval rearing tubs (62 × 53 × 28 cm) containing 1·5 kg of larval food mixed with 5 l of water. The food was a mixture of alfalfa, brewers' dried grain and soft wheat bran (the standard CSMA medium). The pupae were removed 7 days after the introduction of the eggs and each tub yielded approximately 20 000 pupae. The separation of these from the medium involved their flotation by the addition of water, removal by sieving and subsequent air-drying.

Two methods have been devised for sex separation in house flies. One (Bailey *et al.*, 1970) is applicable to adult flies and is based on a difference in weight. Anaesthetized flies are put into a inverted U-tube up one arm of which a stream of air is forced. By regulating the flow of air the lighter males can be blown over the bend in the tube and collected. At five days old the weight difference is sufficient for 97% of those insects coming out of the apparatus to be male and 94% of those remaining to be female. The other method is one used for sexing pupae (Whitten, 1969) and involves the incorporation of a sex-limited marker affecting puparium colour. The mutant is in fact an autosomal recessive made sex-limited by crosses involving a male-linked translocation. The net result is that the male has a brown pupa and the female a black, and the consequent difference in their ability to transmit light is used to separate them. Mixed pupae are passed over a transmitted light-source and automatically sorted. Some 7 000 an hour can be dealt with. A natural sex-limited marker producing a black patch in the thorax of mature female larvae of *An. gambiae* species A and species B is a common mutant which might similarly be used. In the experiment already referred to (Davidson *et al.*, 1970) regular inspection of hybrid rearing trays was made for such larvae which were hand-removed on the happily-rare occasions they appeared.

In complete contrast to the prolific breeders so far considered, the tsetse fly with its single offspring produced every nine days or so, presents considerable difficulties from a mass-production point of view and yet this insect, because of its low natural population levels and relative lack of density-dependent compensation, is considered one of the most suited for control by the sterile insect technique (Knipling, 1967). Nash *et al.* (1971) describe in detail the methods used in England and in France for rearing four species of *Glossina: G. morsitans* from Rhodesia, *G. austeni* from Zanzibar, *G. tachinoides* from Chad and *G. fuscipes* from the Central African Republic. All four are kept in Paris; only the first two in Langford in Bristol. The adults are held in netting cages, the mesh of which is large enough to allow born larvae to drop through. They pupate shortly afterwards and are held in sand (Langford) or tissue paper (Paris) until the adult emerges. This takes an average of 33 days in the case of female *G. austeni.* Very high insemination rates are achieved at Langford by tubing a three-day-old fed female with a single male at least 15 days old. Feeding in Paris is entirely on lop-eared rabbits with up to 1 000 flies being allowed to feed on one rabbit in the course of one day. At Langford most of the routine feeding is on goats, with similar numbers per day, though some is on rabbits but here only 160 bites a day are allowed. Rabbit blood produces a significantly higher yield of pupae than goat blood—14·2 pupae per female in *G. austeni* and 10·6 pupae per female in *G. morsitans* as compared with 9·7 and 7·5 from goat feeding. That little change had occurred in the insects through colonization over a period of two years was shown when males of the Langford colony of *G. morsitans* were taken to Rhodesia and released. They dispersed as well as native flies. In addition wild flies were sent from Rhodesia to Langford and found to mate readily with their colony relatives.

Considerations of life-tables and age-specific fecundities have led Jordan and Curtis (1968) to conclude that a colony of 36 000 adult flies of *G. austeni* would be needed to produce an output of 10 000 adult males per week, if feeding were done on lop-eared rabbits. It is calculated that to maintain a colony of this size would require 675 rabbits, if these were not to be irreparably damaged. The same authors (1972) using the same considerations calculate that the weekly output of males from a colony of the shorter-lived *G. morsitans morsitans* could be 18–25% of the total adult stock, a figure slightly lower than the previous one calculated for *G. austeni* (28%).

Among other species of medical importance starting to be reared on a large scale is the reduviid bug, *Rhodnius prolixus*, a vector of Chagas disease in South America. Gardiner and Maddrell (1972) feeding the insects on sheep have succeeded in producing 7 000 bugs per week.

II. Sterilization

The ultimate effect of the exposure of insects to sources of irradiation or to chemosterilants depends on the degree of exposure and the stage of development of the germ cells at the time of exposure. A high degree of exposure can result in somatic damage and developmental abnormalities affecting subsequent behaviour and longevity. The exposure of germ cells before gametogenesis can result in a complete prevention of the formation of gametes and hence absolute sterility. Usually, however, this is also accompanied by a loss in competitiveness with untreated insects. What is aimed at is the exposure of stages in the life history where gametogenesis, particularly spermatogenesis, has already begun and a degree of exposure producing no somatic damage and as little effect on competitiveness as possible, but still resulting in a high degree of sterility. Under these conditions the main cause of sterility is found to be not the non-production of gametes or their inactivation, though these may occur as well, but the occurrence of dominant lethal mutations in the gametes which show their lethal effect in subsequent zygotic development, after fertilization has taken place in fact. The subject has been reviewed in some detail by LaChance (1967), Proverbs (1969) and Smith and von Borstel (1972).

Some of the most convincing evidence that dominant lethals are involved comes from studies on the parthenogenetic wasp, *Habrobracon juglandis*. In this insect, unfertilized eggs give rise to haploid males. If mated with irradiated males, however, eggs do not hatch. Further, if the female is sterilized at such a dose that no hatching occurs and she is then subsequently fertilized by an unsterilized male, normal haploid males of the paternal phenotype are produced. In other words, the cytoplasm of the

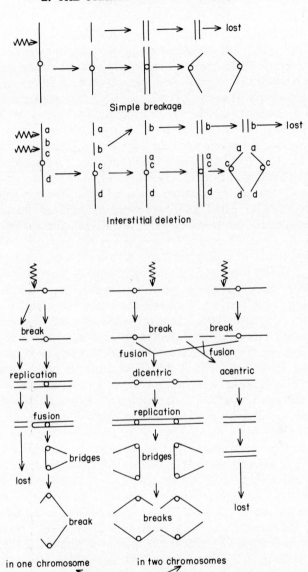

Fig. 1   Chromosome breakage possibilities.

irradiated egg remains undamaged though the nucleus dies and the normal paternal spermatozoon nucleus develops in its place.

Cytological studies have shown that these dominant lethal mutations are not point mutations but rather chromosome breakages which lead to abnormalities in subsequent nuclear divisons. Some of the possibilities are given by Curtis (1971b), (Fig. 1 is a modification of Fig. 1 from this publication). They include single terminal or interstitial breakages and losses of acentric parts, breakage followed by fusion of the sister chromatids still retaining the centromere and subsequent bridge formation or the more complicated fusion of broken chromosomes to produce acentric pieces and dicentrics (with two centromeres) which latter give rise to more bridges. The net result is genetic imbalance due to depletion. There is also some evidence that chromosomal bridges slow down nuclear divisions so that chromosomal and cytoplasmic changes are put out of phase (Smith and von Borstel, 1972).

As already stated, sterilization by irradiation and by chemicals undoubtedly has other effects as well as producing dominant lethal mutations. Aspermia in males can result through two causes. An unlikely one is the treatment at a stage before spermatogenesis has begun or when it is in its earliest phases. In most insects this would involve treatment at a very early stage in the life history as spermatogenesis may be well advanced even before the pupal stage, when as we shall see, for other reasons, treatment is most likely to be given. The most likely cause of aspermia is the killing of those germ cells still in the early stages of spermatocytes and spermatagonia and the depletion of mature spermatozoa.

Very high irradiation doses or chemical concentrations can result in sperm inactivation but such are well above those normally given and those inducing 99% dominant lethal mutations. In *H. juglandis* for example, 5 000–10 000 r produce 99% dominant lethals while 142 000–153 000 r are required to produce sperm inactivation. Similar figures for *Ae. aegypti* are 10 000 r and more than 50 000 r. An exposure of *H. juglandis* for $2\frac{1}{2}$ min. to nitrogen mustard aerosol is enough to produce 99% dominant lethals while an exposure for 90 min. will only produce 53% sperm inactivation.

A depression in egg production can also result from exposures to irradiation sources or chemicals. This may arise from damage to the oogonia or to the nurse cells or both, though nurse cells are relatively insensitive when fully differentiated. This may explain the reason why it takes only 10 000 r to depress the fecundity of *Ae. aegypti* 4 h after a blood meal (when oocyte development is just starting) and takes 100 000 r to produce the same effect some 38 h later (when the ovum is nearly mature).

With notable exceptions, e.g., the Mexican fruit fly (*Anastrepha ludens*), even the dose of radiation necessary to produce a high percentage of dominant lethals may often be accompanied by reductions in mating ability, competitiveness or longevity and this is more often the case than with mutagenic chemicals. As previously stated the aim with both agents is to find the dose or concentration giving the maximum frequency of dominant lethals with minimum effect on post-treatment performance. This frequency need not be 100%, though obviously the higher the better, nor need the female be one which mates only once, as was originally thought. A female which mates many times may ultimately acquire a mixture of dominant-lethal-bearing and normal spermatozoa and then its fertility will depend on the competitiveness of spermatozoa rather than the mating competitiveness of the sterile males.

The stage at which induced sterilization is carried out is largely dictated by considerations of ease of handling during exposure and subsequent release and of at which stage there is the least harmful effect of the sterilizing procedure on mating ability and longevity. The usually-quiescent pupal stage is normally chosen and the least harmful effect seems to be given just before adult eclosion. In the male at least, at this time, primary spermatocytes and spermatids, the stages in spermatogenesis particularly sensitive to the induction of dominant lethals, are usually present.

It seems to be a common finding that, with immature stages at least, the female is sterilized by a lower irradiation dose than the male but that the opposite is the case with chemosterilants. This means that the irradiation dose required to give the maximum degree of sterility of the males with the least effect on its competitiveness will almost certainly sterilize the female completely and that therefore the accidental release of a small proportion of females through faults in the sex separation technique may be of little consequence. With chemosterilants a proportion of females may be unsterilized or incompletely sterilized and then accidental release may be more serious.

The optimum dose for the production of high percentages of dominant lethals may have little effect on the undifferentiated stages of the male germ cells. Spermatogenesis being a continuous process means that after a time sterilized males will in fact recover some of their fertility. This does not usually occur, however, until well after the mating efficiency peak is passed.

III. Release

The time and place of release are bound up with considerations of the normal habits of the insect concerned, the usual aim being to produce a good intermingling of release and wild populations. Adult emergence from released pupae should ideally coincide with the emergence of wild adults and must not occur at some time of the day dangerous to the survival of the adult. A change in emergence time may occur as a result of adaptation to mass-rearing conditions. This must be guarded against. Time has also to be considered in relation to season and abundance of the wild population. This will be dealt with in the later section of this chapter on dynamics, as will be questions of the size and frequency of releases which will of course have to be related to the wild population size and to the relative competitiveness of the sterile insects.

Whether one sex or both should be released may be dictated in the first place by whether or not an easily-operated sex-separation technique is available for the particular species concerned. We have already seen that such techniques do exist for some insects. Secondly a difference in the economic or medical importance of the sexes may determine a policy of single sex release. With the blood-sucking human-disease-carrying mosquitoes, for example, it is only the female which is the biter and vector and a mass release of this sex might lead to an increase, albeit temporarily, of the disease as well as to objections from the human population to increased biting nuisance. Though the female screw-worm is responsible for the damage to domestic animals it was released along with the males because no mass-sexing mechanism existed. Fortunately the radiation dose used stopped female oviposition completely.

Whether the insect mates more than once and whether there is a difference between the sexes in this characteristic are of vital importance dynamically. For example, where the female only mates once (i.e. is monogamous) and the male is capable of mating many times (i.e. is polygamous) the release of sterile males will be more efficient than the release of sterile females, though the release of both sexes should produce more sterility than the release of one sex.

Some confusion exists over the meaning of the terms monogamy and polygamy. As pointed out by Jones (1973) the suffix does not denote a particular sex but merely "marriage partner" or mate. Strictly speaking a monogamous female is monandrous and a monogamous male monogynous but these terms are rarely used. In addition the biological interpretation of the term polygamy varies and it may be useful to define the various aspects of the process of mating as we understand it in insects. Mating usually involves some sort of courtship between male and female followed by the act of connexion of the genitalia—the act of copulation. This is normally followed by the passage of spermatozoa from the male into the female—insemination—and the storage of these gametes until they are required for fertilization of the eggs which may be hours, days, weeks or even months after insemination. The confusion arises when every stage of copulation is taken to imply simultaneous insemination. That this is not necessarily the case is shown very clearly in *Ae. aegypti*.

George (1967) was able to show that the fertility of females of this species mated by irradiated males was not increased by subsequently caging them with normal males and that females mated first with normal males and then with irradiated males lost none of their fertility. He concluded that the spermatozoa received by the female at the first mating have an advantage over the spermatozoa subsequently acquired. It is now known that this is not the correct explanation and that in fact in this species, although both sexes copulate at frequent intervals, insemination only normally occurs once in the female's lifetime. Once inseminated she is normally refractory to further insemination though not to further participation in the act of copulation. Multiple insemination can occur but is a rare event happening only when several males copulate with a female within a short time (VandeHey and Craig, 1958). Craig (1967) showed by implanting various male tissues into the thorax of virgin females that he could make them refractory when male accessory gland tissue was the tissue implanted. When such females were caged with an excess of males none became inseminated if at least four hours were allowed between implantation and exposure to males. Investigations with *C. p. pipiens*, *An. quadrimaculatus*, *An. stephensi* and nine other species of *Aedes* showed the same mechanism operating and even some interspecific effect was seen when the accessory glands of one species were implanted into the females of another species (Craig, 1970). Fuchs *et al.* (1968) later extracted the hormone principle involved from the accessory glands of *Ae. aegypti* and the substance has now been patented under the name "matrone" (Craig and Fuchs, 1969). According to Craig (1970) one male *Ae. aegypti* contains enough matrone to sterilize 80 females. The actual nature of the refractory state in *Ae. aegypti* has been described by Gwadz *et al.* (1971). Females already inseminated go through the act of copulation but position the vaginal orifice guarded by the cerci in such a direction as to prevent the entry of the aedeagus of the male.

Female monogamy is also apparent in species of the *An. gambiae* complex. Dieldrin-susceptible species A females allowed to mate with their own males were later offered dieldrin-resistant males and most (though not all) produced susceptible offspring (Goma, 1963). The most conclusive proof in this species complex has come from experiments carried out by Bryan (1968, 1972) using sterile males produced by crossing member species. Working first with males derived from a cross between species B males and *An. melas* females and possessing well-developed accessory glands, Bryan (1968) showed that when females of species B which had previously laid fertile eggs were artificially mated (using the technique of Baker *et al.*, 1962) with the sterile males they continued to lay fertile eggs. Virgin females of this species readily develop eggs if they are fed on blood but seldom oviposit unless they go through the act of copulation. Such females left with sterile males and successfully ovipositing sterile eggs were then force-mated with fertile males and continued to lay sterile eggs. This apparent absolute refractoriness to the effect of a second mating was even more marked than in *Ae. aegypti* where forced-mating by a similar technique quite often resulted in the second mating taking effect (Craig, 1967). Some of the crosses in the *An. gambiae* complex produce sterile males with very reduced accessory glands. Using two such crosses (between *An. melas* male and species A female and between *An. merus* male and species A female) Bryan (1972) was not able to produce refractoriness to a subsequent mating, thereby implicating the accessory gland. The second function of accessory gland substance in stimulating oviposition originally suggested by Leahy and Craig(1965), was confirmed in these experiments. Females mated with males with reduced accessory glands rarely oviposited while those mated with fertile males and sterile males with normal-looking accessory glands did so readily.

Riemann and Thorson (1969) are also convinced that female monogamy is the normal state in *Musca domestica* especially if the female has reached full maturity at the time of mating. Here the hormone is apparently produced in the male ejaculatory duct and is responsible as in mosquitoes for both oviposition and loss of receptivity in the female though indications are that "incomplete" amounts can induce oviposition without the loss of receptivity. There is also some evidence that females mated very early in life may mate again later.

Multiple mating (female polygamy) apparently can occur in species of *Glossina*. Dame and Ford (1968) exposed virgin female *G. morsitans* first to sterilized males and then to normal males and produced viable offspring, while Curtis (1968c) doing the same thing with *G. austeni* concluded that spermatozoa from the first mating were predominantly used. Proverbs (1962a, b) on the other hand, working with the codling moth (*Laspeyresia pomonella*) showed that a female mated first with an irradiated male and later with a normal male showed a higher fertility than one mated only to an irradiated male, the implication being that the irradiated spermatozoa competed poorly with the normal.

To what extent polygamy in other insects, e.g., fruit flies, involves the act of copulation alone or to what extent it involves multiple insemination seems largely uncertain. Multiple matings in the broadest sense may be more a characteristic of laboratory colonies than of normal wild behaviour. Female navel orangeworms (*Paramyelois transitella*) and codling moths (*L. pomonella*) are both apparently polygamous in the laboratory and yet the former was equally well controlled by the release of sterile females as by the release of sterile males and better controlled by the release of both sexes than by the release of one sex alone (Husseiny and Madsen, 1964), while sterile males alone proved more effective against the codling moth. This could be taken as implying that the former species is functionally polygamous in nature and the latter is in reality monogamous (Proverbs and Newton, 1962).

There seems little doubt that male polygamy is the rule rather than the exception in most insects. From experiments in which males are caged with a large excess of females, the upper limit of male mating ability has been determined as somewhere in the region of 10 inseminations per male in *Lucilia cuprina* (Whitten and Taylor, 1970), 4 in *An. gambiae* species A (Cuellar *et al.*, 1970) and 5 in *G. morsitans* (Dame and Ford, 1968).

Thus the decision on whether to release both sexes of insects sterilized by irradiation or chemosterilants or only one and which one, will depend, among other things, on the mating ability of each sex and in the case of the males, whether sterility is due to aspermia or sperm inactivation or to dominant lethals. If a female is truly polygamic and mates with both an aspermic or sperm-inactive male and a normal male she will be fully fertile. If she is truly polygamic and mates with both a sterilized male producing spermatozoa carrying dominant lethals and a normal male, how fertile or sterile she will be will depend on actual sperm competitiveness. With these considerations in mind Ailam and Galun (1967) calculate that providing the number of available male matings in the population exceeds that of female matings the number of fertile eggs laid will be independent of the number of sterile females while if the reverse is the case, then the net sterilizing effect will be governed by the mating habit of the female. If the female is monogamous, sterility increases with an increase in sterile females and is independent of the number of sterile males. If she is truly polygamous, sterility increases when either the number of sterile females or sterile males increases.

Whitten and Taylor (1970) having a workable sex-separation for *L. cuprina* and knowing that the optimum irradiation dosage with the least effect on subsequent performance differed considerably between male (7 000 r) and female (3 000 r) considered the possibility of releasing males in one area and females in another. On the basis that the upper limit of male mating ability was 10 times, a release of 100:1 ratio of sterile to normal monogamous females should produce a similar effect to a release of a 10:1 ratio of sterile to normal males. If the females were polygamous a lower ratio than 100:1 would be required and these authors set about selecting for this characteristic. After selecting for 10 generations, the proportion of females mating twice had in fact increased from 1–5% to 30%, and so Whitten (1971c) went on to consider the effect of the release of mixtures of polygamous and monogamous sterile females. To produce 90% control he calculated that he would need to release between 15 and 100 sterilized females for each wild male and female. Thus the small additional control from a mass-rearing facility producing a normal sex-ratio would not warrant the separate release of what is after all the economically important sex.

These are largely theoretical considerations. Most would need experimental field release programmes to confirm their practical validity. Of increasing practical concern is the possible production, through mass-rearing and sterilization, of insects showing behavioural differences in dispersal and mating preferences which on release as mixed sexes may show limited dispersal and a tendency to self-mate rather than seek out the wild individuals. As Cuellar (1973a) points out, such behaviour would lead to a wasting of part of the sterile insect mating capacity. From all these considerations it can be concluded that though the release of the male sex is not imperative for success in the sterile-insect control technique, it is preferable in most instances.

IV. Dynamics

The idea of using deliberately sterilized insects to control their own kind came from Knipling, a United States Department of Agriculture entomologist originally concerned with the control of livestock pests. In a number of publications, but perhaps most thoroughly in one of 1967, he compares the efficiency of the sterile insect method with conventional killing methods, e.g., insecticides. His models are of the simplest kind of an initial insect population of one million individuals able to reproduce themselves at a constant rate of five times in every generation in an environment that will support a maximum of 125 million insects. If such a population were subject to a 90% kill in each generation by the application of an insecticide (or a pathogen or a predator or by some other means such as breeding place reduction) then the population would be reduced to less than two individuals in 19 generations—the population would in fact be halved in each generation. If, on the other hand, 9 million equally competitive sterile insects were released into the same population in each generation, eradication would be achieved after five releases (45 million insects) because the sterilizing influence would increase in each generation as the proportion of sterile to normal insects increased. At the second release this proportion would be 18:1, at the third 68:1, at the fourth 944:1 and at the fifth 180 000:1. The sterilizing influence would therefore increase from 90% after the first release to more than 94% after the second, more than 98% after the third and more than 99% after the fourth. This in no way allows for a possible persistence of sterile insects from one generation to the next which may well occur and then the proportion of sterile to normal insects would be even higher.

The basic difference between the two methods of control is that while the first is efficient when the population is large but less so when it is small, the sterile-insect method is relatively inefficient against a large population, but very efficient against a small one. Thus a first combined attack with insecticide and sterile-insect release followed by sterile-insect releases alone, and at one-tenth of the numbers required when only the sterile-insect method is used, will produce eradication in one generation less and using only 3 600 000 sterile insects (Table I).

TABLE I   Relative trends of hypothetical insect populations subjected to control by insecticide, by sterile-insect releases and by a combination of the two methods.

Generation	Uncontrolled population	Insecticide treatment (90% kill in each generation)	Sterile-insect releases (9 000 000 in each generation)	Initial application of insecticide (giving 90% kill) combined with sterile-insect release (900 000) followed by a similar sterile-insect release in each subsequent generation
Parent	1 000 000	1 000 000	1 000 000	1 000 000
1	5 000 000	500 000	500 000	50 000
2	25 000 000	250 000	131 580	13 160
3	125 000 000	125 000	9 535	955
4	125 000 000	62 500	50	1
5	125 000 000	31 250	0	
6	125 000 000	15 625		
7	125 000 000	7 812		
8	125 000 000	3 906		
9	125 000 000	1 953		
10	125 000 000	976		
11	125 000 000	488		
12	125 000 000	244		
13	125 000 000	122		
14	125 000 000	61		
15	125 000 000	31		
16	125 000 000	15		
17	125 000 000	8		
18	125 000 000	4		
19	125 000 000	2		
20	125 000 000	1		

Even more efficient than the mass-rearing, sterilization and release of sterile insects would be the introduction of a sterilizing agent into a wild population. Knipling (1967) considers such an agent affecting 90% of the population in each generation. Final eradication takes the same number of generations as in the case of sterile insect releases, but reductions per generation are greater, providing both sexes are fully sterilized : one million → 50 000 → 2 500 → 125 → 5 → 1. This method eliminates the need of a mass-rearing facility and could be applied using chemosterilants.

Because of the toxic hazards associated with most of these chemicals, however, they would have to be applied in controlled situations not presenting contamination dangers, and one essential for their efficiency would be means of attracting wild insects to them. Sex and food attractants or physical attractants such as light and sound sources are possibilities but no such attractant–chemosterilant combination has yet been successfully demonstrated in the field.

Knipling's simple models are easily criticized. He himself is well aware of their short-comings with respect to reproduction rate and mating efficiency in particular and comments that no account is taken of pressure effects of large scale releases or of possible survival of previously released sterile insects at the time of subsequent releases. What he aims to do, and nothing more, is to point out the potential of the technique and compare it with conventional methods.

Geier (1969) stresses the artificiality of a constant reproduction rate. The reproduction rate is a direct function of population size and available food and breeding area. When the population is low and food and breeding area high, the reproductive rate will be high; when the population is high and the environment nearly filled to capacity, this rate will be low. Using a population of 1 000, an environmental capacity of 100 000 and a declining reproduction rate starting at twenty times and decreasing to three times, one and a half, etc., he estimates eight generations to produce the maximum population size. Flooding a population of this kind with 9 000 sterile insects every generation will barely affect the rate of increase; in eight generations the population size will be more than 80 000 and will remain on the increase. However, if a high enough initial release is made (just below 38 500 sterile insects) and proportionally lower releases in subsequent generations (24 339; 9 342; 906; 202; 47 and 11) eradication can be achieved in seven generations using a total of 73 341 sterile insects. Constant high releases can also produce eradication but require higher total numbers of insects. 20 000 per generation achieves eradication in nine generations taking a total of 180 000 insects; 100 000 per generation takes two generations and 200 000 insects, while eradication can be produced by a single release but necessitates 400 500 insects.

Inevitably the sterile insect technique has attracted the mathematically-minded biologists and others to construct models from the component parameters of insect population production and, utilizing the computer, to determine the relative importance of each of these parameters in populations challenged by sterile releases. Such exercises are useful in highlighting the really important parameters and in stimulating the biologist to produce accurate estimates of them. They are also extremely helpful in predicting needs and outcomes of actual release programmes.

Cuellar (1969a, b) has produced such a model based on parameters applicable to *An. gambiae* and considers what he terms the critical level of interference required for eradication. As we shall see when we come to consider hybrid sterility, member species of the *An. gambiae* complex have not only a high biotic potential in terms of egg-batch size, frequency of oviposition and female longevity but also a notorious reproduction rate which reaches a peak shortly after seasonal rainfall provides breeding places in the form of newly-created predator-less pools. The parameters having the greatest effect on population size are the average number of eggs laid by a female in her life-time—a function of egg batch size and longevity—and the probability of an egg becoming an adult (V). The effect of residual insecticides applied to human dwellings (the usual method of malaria control) is on female longevity while the effect of sterile insect release would be on the overall fertility index—a function of the probability of fertile female meeting fertile male and the relative mating abilities of fertile and sterile individuals.

Cuellar (1969b) considers a population with a daily output of 10 000 individuals and the following parameters operating:

(1) 100 eggs per oviposition;
(2) the maximum probability of an egg becoming an adult as 0.9;
(3) an aquatic cycle of seven days;
(4) an oviposition frequency of every two days but four days before the first oviposition;
(5) a maximum possible length of life of a female of 30 days;
(6) an overall fertility index of 1·0 in wild individuals;
(7) an adult female daily survival rate before treatment of 0·9.

An insecticidal treatment of such a population changing the probability of survival from 0·9 to 0·2 would produce eradication in nine weeks if the value of V is 0·01761 at the beginning and allowed to increase to 0·9. A less efficient insecticidal application only resulting in a change of female probability of survival from 0·9 to 0·3 would not produce eradication no matter how long it was continued, if the same initial value of V applies and expansion to 0·9 is allowed. With the same population the daily release of 260 000 sterile males would eradicate in 19 weeks while the daily release of 240 000 would never do so. A daily release of 300 000 would eradicate in 13 weeks and actually involve the release of fewer total numbers.

Conway (1970) working quite independently of Cuellar has also produced a computerized anopheline model. By simulating a strongly density-dependent population with a mean size of 10 650 adults and challenging it when stabilized with 6 releases of 250 000 sterile males every two days, he shows that no eradication is achieved. However, if the initial release is combined with an insecticidal application producing 90% mortality among the aquatic stages and a 95% residual effect then the 5 further releases of sterile males would result in population elimination. With a more weakly-buffered population of the same size, eradication can be achieved by the release of sterile males alone.

One very simple point emerges from these mathematical simulations. If the female only mates once and the state of her offspring (viable or non-viable) is determined for life by this one mating (and this seems to be the case with many insects), then sterile male releases will have to be continued for at least the maximum length of life of the female, plus the length of an aquatic cycle and the time from adult emergence to mating receptivity to achieve eradication. In other words, females already inseminated before releases start will go on adding viable individuals to the population for the rest of their lives irrespective of how many sterile males are present. The persistence of these fertilized females in a naturally declining population challenged with sterile males can result in very different proportions of sterile to normal females from those calculated for a stable population (Cuellar, 1970). In a stable population with a probability of daily survival of $0\cdot9$ challenged with sterile males at a ratio of $2\cdot5$ to every normal male the expected composition of the female element on the thirty-sixth day after the start of releases would be 71% sterile and 29% normal. In a uniformly declining population where the population size at the end of this time is postulated as 5% of the initial size, the proportion would be only 30% sterile and 70% normal, given the same treatment. The effect of the population decline is to alter the age composition and results in an increase in the proportion of the older age-groups.

More general models have been constructed by Berryman (1967) and Kojima (1971). Bogyo *et al.* (1971) used Berryman's model to calculate minimum ratios of sterile to fertile males necessary to produce population elimination in five generations, varying the parameters one at a time to assess their relative importance. Three mating conditions were considered: (1) where the female mates several times throughout her life and the spermatozoa she acquires are used indiscriminately; (2) where the female is polygamous but mating ceases after the first oviposition and the spermatozoa are again used indiscriminately; (3) where the female only mates once, or, if she is polygamous, only the spermatozoa from the first or last mate are used. The parameters varied were the probability of survival from egg to adult, the proportion of females in the population, the size of the population, sterile male competitiveness and spermatozoa competitiveness. The most crucial parameter proved to be the probability of survival from egg to adult. In every case the first mating condition of female polygamy throughout life demanded the highest proportion of sterile to fertile individuals for eradication and the third condition of only the spermatozoa from one mating being used the lowest.

In Kojima's model more of the parameters are allowed to vary simultaneously and a new parameter introduced. This might be termed the "genetic resistance factor" and allows for improvement in survival rate and fecundity—fertility rate resulting from the selective effect of the control method. It may be remembered that Boesiger (1972) considers this a likelihood. Kojima assumes both male and female are monogamous and concludes that the release of x sterile males alone is more effective than the release of both sexes ($x/2$ males $+ x/2$ females) even when both are sterile.

All these dynamicists agree that the all-important factor dictating success or failure of the sterile-insect technique is the ability of the natural population to respond to deliberate reductions in its fertility by a compensatory increased likelihood of individual survival of those remaining fertile insects. To be able to measure this potential could give some idea of the amount of sterility needed to be introduced to prevent any recovery. Detailed life-table studies on classical ecological lines could give an answer but would take considerable time and observations to produce a meaningful result. Weidhaas (1968) from a study of Knipling's simple models recognized that the rate of increase of a population could be relatively simply estimated in the field by measuring density changes from levels in the generation before the introduction of a known sterility factor to those in the following generation or generations. Thus if densities remained the same before and after a release of equally competitive sterile insects in the proportion of nine sterile to every normal one, then the population must have an ability to increase its numbers by ten times at least. Weidhaas and LaBrecque (1970) applied this principle to the findings from a field attempt to control house flies in Grand Turk Island by the use of chemosterilant (metepa) baits applied to all privies on the island. Flies were counted as they rested on standard grids placed in kitchens and their sterility assessed in samples caught and allowed to oviposit. From these records it was possible to assess the degree of sterility in the parental generation and the reduction in density in the F_1. We know from Knipling that 90% sterility results in the halving of a population with a five-fold reproductive potential. In the housefly population of Grand Turk Island the sterility varied from 60% to 91% and the density reductions varied from 0% to 62%. Relating these figures from one generation to the next resulted in estimations of reproductive potential of from one to eleven times but mostly between four and five times. This was considerably less than the 145 times estimated from laboratory populations held under optimum conditions.

Weidhaas et al. (1972) extended these calculations to the Seahorse Key attempt to eradicate C. p. fatigans by the release of chemosterilized males (Chapter 4) and to control the stable fly (Stomoxys calcitrans) by the release of both sexes sterilized with a chemosterilant in a dairy in Florida (La-Brecque *et al.* 1972a). Reproduction rates varied from less than one to eleven times in *C. p. fatigans* and from 1·1–3·2 times in *S. calcitrans*. Rates of this order would certainly not prohibit consideration of the use of the sterile-insect method of control.

Cuellar and Cooper (1973)[1] have attempted to put the net reproductive rate in mathematical terms with the following formula:

$$\frac{Vp^j bf}{1 - p^k}$$

where V = probability of survival from egg to adult emergence
p = probability of adult female daily survival
j = interval in days between adult female emergence and first oviposition
b = average number of eggs per oviposition
f = proportion of female eggs in the average oviposition
k = interval in days between the first and subsequent ovipositions

The probability of survival from egg to adult emergence when the reproductive rate is unity, i.e., when the population is in a state of equilibrium (Veq)

$$= \frac{1 - p^k}{b^f p^j}$$

The net reproductive rate can then be calculated if the maximum value of V (Vmx) is known from:

$$\frac{Vmx}{Veq}$$

A maximum value of 0·8 was suggested for *C. p. fatigans* by Weidhaas *et al.* (1971) and the various parameters in the Seahorse Key field trial were estimated as:

$p = 0·75$
$b = 180$
$f = 0·5$
$j = 8$
$k = 4$

From these figures $Veq = 0·10$ and the net reproductive rate is 10·7 times, a figure not unlike those calculated from considerations of sterility proportion and change in density.

Field-cage experiments with the same species (Patterson *et al.*, 1972b), on the other hand, indicated reproduction rates greater than 29 times and less than 99 times from the release ratios of sterile males and the proportion of egg-rafts found to be sterile. Parameters given for the Delhi area

[1] Cuellar, C. B. and Cooper, A. World Health Organization mimeographed document WHO/VBC/73.460.

by Yasuno (1972) indicate higher rates too:

$$p = 0.8$$
$$j = 4$$
$$k = 3$$
$$b = 250$$
$$f = 0.5$$

Such values give a Veq of 0.0095 and if the maximum value of V is taken as 0.8 the reproductive rate could be as high as 84 times, though it should be pointed out that Yasuno (1972) considers Vmx to be only 0.09 and the reproductive rate only 9.5 times. It will be noted that the values of j and k given by Yasuno (1972) are lower than those given by Weidhaas et al. (1971) and in fact might be more representative of "height of season" conditions. If these values are substituted for those estimated in Seahorse Key, the net reproductive rate in this area becomes 38.1 times instead of 10.7 times. Cuellar and Cooper (1973)[1] think the oviposition cycle (j) may be even as low as two days and the probability of adult survival (p) as high as 0.9 or 0.95. Then the maximum reproductive rate in Seahorse Key could be as high as 249 times ($p = 0.9$) or even 610 times ($p = 0.95$). Such values of j, k and p are known in anopheline mosquitoes.

Thus the possibility exists of much higher rates of increase than those estimated by Weidhaas et al. (1972), though it must be recognized that a great deal depends on what value is given to the maximum value of survival from egg to adult. The figure of 0.8 may seldom be realized in the field. On the other hand, the field estimates depend on the reliability of sterility and density observations. The fact remains that the impact of sterile insect releases will be greater against populations with a low net reproductive rate than against those with a high one, and the general consensus of opinion is that the greatest chance of success with such a method will be when the population is naturally declining. This natural decline may be due to changes in climatic conditions reducing survival and lengthening generation time or to changes in availability of breeding environment, both food and site, or more likely a combination of all these things. The additional impact of a degree of sterility will reduce abundance and there may be temporary compensatory increases in reproduction rates but these will certainly be smaller than in situations where breeding conditions are expanding and climatic conditions optimal.

LaBrecque (1972) considers the pros and cons of releasing sterile insects against declining and expanding populations and makes the point that factors producing population decline may also affect the released insects. This is a very valid point as the insects released are likely to have been reared under conditions resembling those that allow natural population expansion and may not be able to seek out the few remaining wild

[1] Cuellar, C. B. and Cooper, A. World Health Organization mimeographed document WHO/VBC/73.460.

individuals. An extreme example would be when the few remaining females (already normally fertilized) go into hibernation or aestivation. No amount of sterile insect release could have any effect on such individuals during this period. Whilst admitting that the numbers of sterile insects required to combat a rapidly expanding population may be uneconomical he suggests a start of releases when populations are just beginning to increase and the continued release of enough insects to contain population growth during the season of greatest potential abundance. Then when the adverse climate and other effects make their appearance, the population is no longer resilient enough to overcome them. This argument could only be relevant where maximum rates of increase are of the order determined by Weidhaas *et al.* (1972), up to some ten times, and could never apply to the kind of reproductive increase calculated by Cuellar and Cooper (1973).[1]

One dynamic aspect so far only considered by Cuellar (1973c) is the difference in behaviour of the two sexes in the wild, leading to an accumulation of one sex in the mating situation. This has been considered in relation to mosquitoes, but may also be applicable to other inects. In mosquitoes the male is not necessarily short-lived. A measurement by the mark–release–recapture technique carried out in Tanzania with *An. gambiae* species A showed the probability of survival of males was 0·82 while that of females was 0·84 (Gillies, 1961). The monogamous female, characteristic of mosquitoes, leaves the "mating pool" after insemination and does not return. This combination leads to an accumulation of males in the "mating pool". Cuellar calculates 4·43 males to every female where the female is receptive the night following the night of emergence and the male capable of mating one night later, providing all virgin females of this age are fertilized on one occasion. Thus the general assumption that the number of sexually active fertile males in the "mating pool" is the same as the numbers emerged on the previous day may lead to an underestimate of sterile male requirements, at least initially.

What is obvious from all these largely theoretical considerations is the absolute necesssity of comprehensive ecological studies being made of each insect where control by sterile-insect release is contemplated. Population size determinations are essential for estimates of sterile release requirements. A detailed knowledge of seasonal changes in abundance must be at hand to aid decisions on the best time to release. The flight range of the species to be controlled must be known in planning the extent of the area where releases will have to be carried out and the maximum length of life of a female will need to be estimated to assess the minimum period of time that releases will have to be continued. Above all, efforts should be made to determine those periods in which density dependence operates with the least efficiency.

[1] Cuellar, C. B. and Cooper, A. World Health Organization mimeographed document WHO/VBC/73.460.

3. Sterilization by Irradiation

Though insect sterilization can be brought about by exposure to alpha and beta particles and to X-rays and neutrons, gamma radiation sources are the most convenient means of achieving this end. The most common source is ^{60}Co with its half-life of six years. ^{137}Cs gamma-irradiators have a much longer half-life (thirty years) but longer exposures are needed to give equivalent radiation doses.

The establishment of optimum dose rate for each sex of a particular insect species involves the exposure of a particular age and stage to a series of exposures and the subsequent mating to the opposite untreated sex, to determine from egg hatches the percentage of dominant lethals induced. LaChance (1967) reviews some of the experimental work done on these lines. Males are treated when mature sperm are present and then mated to virgin untreated females. An initial linear relationship between dose and frequency of dominant lethals is followed by a levelling-off and an area where quite large increases in dose produce only small increases in dominant lethals. It is precisely this area that is of crucial importance from a practical point of view and where a decision has to be made on whether to sacrifice competitiveness and longevity for very high degrees of sterility. It is usual to accept the point where the levelling-off starts, when the incidence of dominant lethals will be less than 100% but is usually above 90%. The doses giving equivalent dominant lethal frequencies vary from one species to another. For example, to produce a 90% frequency requires an exposure to 5 000 r of males of the screw-worm (*Cochliomyia hominivorax*) and eye-gnat (*Hippelates pusio*), to 8 000 r of the honey bee (*Apis mellifera*) and reduviid bug *Rhodnius prolixus*, and to 35 000 r of the codling moth (*Laspeyresia pomonella*). The result of the exposure of females depends very much on the developmental state of the oocytes at the time of exposure. In *C. hominivorax* for example, when the oocyte nucleus is in the resting stage (prophase I) a sigmoid relationship between dose and dominant lethal frequency is apparent with maximum effect after 5 000 r and up to 8 000 r when the frequency is of the order of 70%. After this an increase even to 14 000 r does not produce more than 80% lethals. With more mature oocytes (anaphase I) the relationship parallels that found with mature spermatozoa—a linear relationship between 500 and 2 500 r and then a levelling-off with some 85% dominant lethals being induced from exposures to 3 500 r.

Various external factors can affect the outcome of irradiation, particularly the atmosphere in which exposure is made. It takes 5 500 to 6 200 r to produce complete sterility in *C. hominivorax* when pupae are exposed in air and only 4 500 r when they are exposed in a 50–50 mixture of carbon dioxide and air (LaChance, 1963). Pure carbon dioxide and pure nitrogen on the other hand are radiation protectors (LaChance, 1962). Fractionating the dose usually reduces the ultimate sterility. Grosch and Sullivan (1954) for example, found that the delivery of 5 000 r to *Habrobracon juglandis* over 50 min. did not achieve full sterility, while the same dose given in eight min. did so. The temperature before, during and after exposure can also have marked effects on the final outcome.

I. Livestock Pests

A. The Screw-worm

Beyond a doubt the most impressive practical application of the sterile-insect method of control has been with the screw-worm (*C. hominivorax*) sterilized by gamma-rays. This muscid fly is a parasite of warm-blooded animals from the southern parts of the United States to South America and certain West Indian islands. The female lays her eggs in skin lesions where the larvae hatch and feed for five to six days before falling to the ground and pupating in the soil. As well as economic loss through hide and carcass damage, death from heavy infestations may occur especially when the navel of newborn animals is infested. Prior to 1958 the estimated annual loss to the livestock industry in the United States of America was $120 000 000.

The first successful attempt to control the screw-worm by the release of irradiated flies was made in 1952 on a small island, Sanibel, two miles off the west coast of Florida. This is an island, sixteen square miles in area, where in winter the screw-worm population was estimated to be some ten flies per square mile. 200 irradiated flies (100 of each sex, there being no simple sex-separation technique) were released in each square mile once a week. After 2 weeks 80% of the egg masses sampled proved to be sterile. After 3 months the fly population was almost zero but was never eradicated because of the island's proximity to the uncontrolled mainland.

The next field trial was carried out on the island of Curaçao off the north coast of South America, some forty miles from the nearest other land. Here 200 sterilized flies a week were not enough and the number had to be increased to 800. Over an area of 170 square miles this meant a weekly release of 136 000 flies. Eradication was achieved in nine weeks (Baumhover *et al.*, 1955).

Before 1933 the screw-worm was confined in the United States to those south-western areas near to the Mexican border. It was accidentally introduced by way of infested cattle imported from Texas into Georgia and from there it spread to Florida where it was able to overwinter. In 1957 a start was made to eradicate the insect from the south-eastern states. A mass-rearing facility was established at Sebring in Florida and produced 50 million flies a week using the methods described in Chapter 2. Five-and-a-half-day-old pupae were irradiated at 7 500 r and adults released shortly after their eclosion, which occurred one or two days after irradiation. This irradiation level stopped female oviposition completely. The quality of the factory-reared and sterilized flies was routinely tested not for mating efficiency but for aggressiveness as judged by the mortality among females caged with males at a 3:1 ratio of males to females. In this Sexual Aggression Test (SAG) a total of 80 flies were held in a cage 15 × 15 × 30 cm for 15 days and mortalities recorded at intervals. Over 80% mortality normally occurred among the females (Baumhover, 1965).

Over Florida and the neighbouring states 800 flies per square mile per week were dropped from aircraft in flight lanes twelve miles apart. These lanes were moved each day so that at the end of a six-day working week the area was covered by two-mile swaths and at the end of two weeks by one mile swaths. Additional releases were also made on known heavily infested areas. After seventeen months and the release of $3 \cdot 5 \times 10^9$ (3 500 000 000) sterilized flies complete eradication was achieved at a cost of $10 600 000 and was estimated to save an annual cost in death and damage of $20 000 000 (Lindquist, 1963; Bushland, 1971).

Control in the south-western states presented a much more difficult enterprise. Here there was no natural isolation, such as was provided by the Florida peninsula, but a continous source of invasion from uncontrolled Mexico along a 1 500-mile border. Though mark–release–recapture experiments had shown that most recaptures were made within 50 miles of release points, one was made up to 180 miles distant, and subsequent invasion evidence even indicated a possibility of 300-mile flights. Thus as well as routine releases of sterile flies over such states as Texas, New Mexico, Arizona, California, Louisiana, Arkansas and Oklahoma, barrier releases had to be maintained up to 350 miles in depth over the Mexican border. A mass-rearing factory with an internal area of 76 000 square feet and employing more than 300 people was set up at Mission, Texas, at a cost of $650 000. It had the capacity to produce 150 million flies a

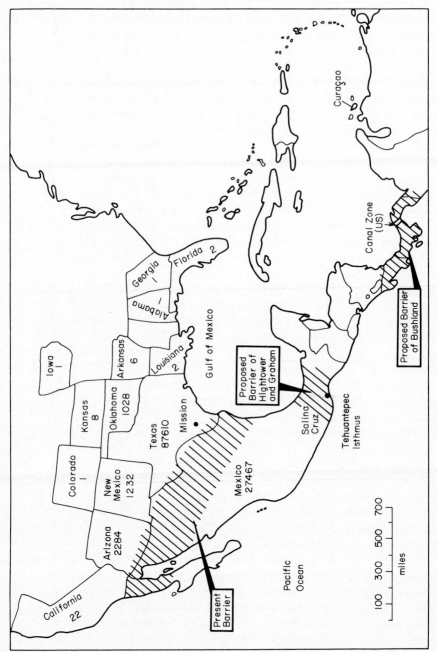

Fig. 2 Map showing the distribution of cases of screw-worm, 1972, and barriers.

week. Releases started in 1962 with an estimated 1.8×10^9 flies. This was increased to 6×10^9 in 1963 and virtual eradication was achieved in Texas and New Mexico by 1964 and Arizona and California by 1965. According to Christenson (1966) the screw-worm infestation cases in Texas stood at 50 779 in 1962. By 1963 they were down to 6 339 and 1964 to 239, though many of the 1964 cases were considered direct invasions from Mexico. According to Bushland (1971) there has been no continually breeding population in Texas since 1964 or anywhere in the United States. Since 1964 barrier releases of 108–125 million flies per week have

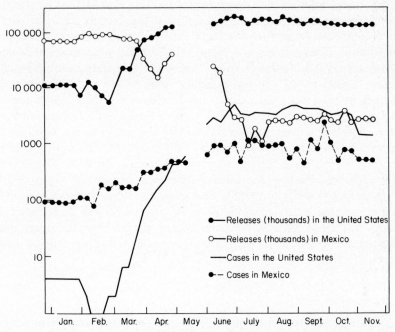

Fig. 3   Cases of screw-worm and releases of sterile flies per week, 1972, in the United States and Mexico. (Data from the Cooperative Economic Insect Report, United States Department of Agriculture).

been made along the Mexican border with concentrated effort on those areas, e.g., watercourses, where cattle and flies are known to congregate. This barrier release programme has been costing in the region of $6 000 000 a year. Total costs of the eradication scheme between 1962 and 1964 have been estimated as $12 200 000 but with an annual saving of some $100 000 000 to the livestock industry (LaBrecque and Keller, 1965; Christenson, 1966; Hightower and Graham, 1968; Bushland, 1971).

This healthy situation was in fact maintained for the best part of seven years, but a major breakdown occurred in 1972. Details of this taken from the weekly Cooperative Economic Insect Report published by the United States Department of Agriculture have been given in *Nature* (Lond., 1973) and show a total of nearly 92 200 cases for most of that year. The accompanying map (Fig. 2) shows the distribution of these cases while the graph (Fig. 3) depicts the numbers of cases in Mexico and the U.S.A. and the sterile fly releases in the two countries month by month. One can only speculate at this juncture as to the reasons for this break-down. Perhaps the mass-rearing of the fly over such long periods and on such artificial diets has changed its behaviour to the extent that it no longer readily mates with its wild counterpart. As long ago as 1968 Hightower and Graham recognized the possibility that the Florida strain of screw-worm might not be suitable for use in the south-western states and started a new strain from collections made on the east and west coasts of Mexico for release on the Mexican border. Perhaps wild material should be con-tinually added to the factory breeding stock to try to eliminate major changes in behaviour. Because of the breakdown a joint agreement has now been drawn up between the United States and Mexican govern-ments whereby a new fly factory will be set up at Salina Cruz in the Tehuantepec Isthmus with a capacity of 300 000 000 flies with the inten-tion of attempted eradication from all territories north of this 140 mile wide isthmus. Such a narrower barrier was originally suggested by Hightower and Graham (1968). As was pointed out by Bushland (1971) an even narrower barrier would be in Panama and Costa Rica. However, eradication from central America where conditions are much more favourable for the all-the-year-round survival of the fly will be much more difficult.

B. Tsetse Flies

Like the screw-worm, the tsetse fly (*Glossina* spp.) normally occurs in low density, and Knipling (1967) is convinced that it could be controlled by the release of sterilized insects. Species of this genus, it will be remember-ed, are primarily transmitters of animal trypanosomiases. Their role as vectors of the human sleeping sickness is comparatively restricted these days. Tsetse flies have a very low reproductive rate, only producing single offspring every nine or ten days. Knipling rates their maximum reproductive capacity as no more than twice. They are notoriously suscep-tible to insecticides and so far no case of resistance has been recorded among them, but attempts to control them by aircraft and ground spray-ing have proved prohibitively expensive. Potts (1958) was the first to show that they could be sterilized by exposing the pupae to gamma rays. Dean *et al.* (1968), working in Rhodesia, exposed wild-caught pupae of *G. morsitans* to a series of irradiation doses, and showed that male flies emerging one week after exposure to 8 000 to 15 000 r were 95–97% sterile, those emerging two weeks after exposure to 9 000 r were 100% sterile as were those emerging three weeks after exposure to 4 000 r. The younger the pupae the higher was the mortality both in pupae and emerging adults. No mortality occurred, however, when a few pupae of known age (17 and 29 days) were exposed to 8 000 r. Male pupae proved more sensitive than female ones, and females emerging from pupae exposed to as little as 1 000 r were completely sterile. Treating the adult male a week after emergence with 12 000 r gave better adult survival than treating the pupae one week before eclosion. Cage competition trials in both small (8 × 8 × 11 in.) and large cages (8 × 12 × 24 ft) at ratios of four sterile males to one normal male to one normal female gave the expected 80% reduction in fertility. More recently Itard (1971) has produced total sterility without affecting fitness, by exposing adult males of *G. morsitans* to 19 000–20 000 r and those of *G. tachinoides* to 15 500–17 000 r from a ^{137}Cs source. Langley and Abasa (1970) working with *G. austeni* showed that 10 000 r applied to pupae which had completed two-thirds of their life, effectively sterilizes males, and that 80% of exposed puparia produce adults which show no effect of the treatment on feeding frequency, meal size, fat body development, longevity and thoracic flight-muscle development. The major stumbling-block to field application of the sterile-insect technique to the control of tsetse flies, is the difficulty of rearing large numbers. This has already been referred to in the previous chapter.

C. Other Livestock Pests

Among the other insect pests of livestock the sterile-insect technique using gamma rays is being considered for torsalo (*Dermatobia hominis*). the cattle grub (*Hypoderma lineatum*), the hornfly (*Haematobium irritans*) and the stable fly (*Stomoxys calcitrans*). The last two of these species can already be reared on a large scale, and preliminary sterilizing doses of radiation have been established for all four (see Control of Livestock Insect Pests by the Sterile-Male Technique, *International Atomic Energy Agency Publication*: STI/PUB/*184* Vienna, and Sterility Principle for Insect Control or Eradication, *International Atomic Energy Agency Publication*: STI/PUB/*265* Vienna, 1971). Eschle has in fact released horn flies in Texas irradiated as pupae with 2 000 r at a ratio of ten sterile to one normal, and shown a reduction in reproduction of 98% six weeks later and a reduction of at least 90% continuing for a further seven weeks.[1] Weidhaas *et al.* have also carried out some releases of *S. calcitrans* irradiated as three to four-day-old pupae with 2 000 r in Florida. 16 000 to 80 000 flies per day were released for five days a week. These flies were

[1] In World Health Organization mimeographed document "Vector Genetics Information Service 1972", VBC/9/73.1.

marked, and recaptures some nine or ten weeks after releases started showed that the ratio of sterile to fertile was in excess of 1:1. Sterilities in females recaptured ranged from 47 to 83%. Laboratory quality control tests of samples of material to be released showed male sterility to be between 98 and 100% and no irradiated females ever oviposited.[1]

II. Agricultural Pests

A. Fruit Flies

Among the agricultural pests, sterilization by irradiation has been applied to the fruit flies more than to any other group, and some small-scale field releases of such sterilized flies have been successful. The earliest attempts were made by personnel of the Hawaiian Fruitfly Investigation Laboratory in Honolulu, using the Mariana Islands as their field sites. One of these, Rota, an island of 34 square miles, had a high population of the oriental fruit fly (*Dacus dorsalis*) and a moderate one of the melon fly (*D. cucurbitae*). First attempts to eradicate *D. dorsalis* by a weekly release of 10 million sterilized flies failed, and it was concluded that the numbers were insufficient (the ratio of sterile to normal flies never exceeded 10:1) and the male not competitive enough (Steiner *et al.*, 1962). Later attempts were made against the melon fly, and combined the use of bait sprays to reduce the initial population by 75%, followed by weekly releases both from the air and from the ground of 8 million sterilized flies. Altogether 257 million flies were released. The pupae from which these flies emerged had been irradiated at 9 500 r in Honolulu. The estimated ratio of sterile to wild flies was 15:1 and complete eradication was achieved in four to six months, the time required for the lapse of four generations (Steiner *et al.*, 1965a). The oriental fruit fly was eventually eradicated from the same island by the method of male annihilation using a combination of attractant (methyl eugenol) and insecticide (naled) (Steiner *et al.*, 1965b). The oriental fruit fly was also eradicated from the island of Guam by the release of 16 million sterilized flies between September 1963 and February 1964, but only after a hurricane had drastically reduced the wild population (Steiner *et al.*, 1970). A few flies, however, were found in the following year, and these were dealt with by the combined use of bait sprays, methyl eugenol and sterile-fly releases. The other islands, Saipan, Tinian and Agiguan, were also cleared of oriental fruit fly, but mainly by the male annihilation method (Steiner *et al.*, 1970). An attempt to eradicate the melon fly from Guam was in progress in 1969 and if it was successful the whole of the Marianas would be free of fruit flies of economic importance (Chambers *et al.*, 1970).

[1] In World Health Organization mimeographed document, "Vector Genetics Information Service 1972", VBC/9/73.1.

More recently attention has been paid to the application of the technique to the Mediterranean fruit fly (*Ceratitis capitata*) both in the new and old worlds. Laboratory experiments have shown the usual relationships between irradiation dose, pupal age and male competitiveness. Hooper (1971), using pupae two days before adult eclosion, showed that 7 000 r produced 95·5% sterility in males and that it took 11 000 r to produce more than 99% sterility. The females were more radiosensitive, and at 3 000 r and above, produced no eggs, or only a few, which never hatched. In general the shorter the time before eclosion when irradiation was carried out the lower the sterility but the higher the competitiveness of the males. 9 000 r applied to pupae two days before eclosion (the regime adopted in most field-release trials) lowered male competitiveness somewhat. No evidence or recovery of fertility of males irradiated at 5 000, 7 000 or 9 000 r was found up to 26 days after irradiation.

Hooper (1970) produces evidence that female polygamy in this species is in fact limited. One mating is sufficient to maintain good egg fertility for three to four weeks apparently. In alternate mating experiments it was found that only 15–21% of females mated with normal males if they had been mated earlier with sterile males, and only 8–12% mated with sterile males if they had mated with normal males previously. A fertile mating following a sterile one changed the egg-hatch from 2–87% while a sterile mating following a fertile one changed the egg-hatch from 100–43%. This suggests that the spermatozoa of irradiated males are not as competitive as those from untreated males or that sterile males transfer fewer spermatozoa.

Hooper and Nadel (1970) describe problems of irradiation and transport prior to the field release of sterilized flies on Procida Island off the west coast of central Italy. A Gammacell 220 irradiator with an exposure chamber of 3 litre capacity was used, but to reduce variance in dose associated with the geometry of the chamber to ±11%, only 0·9 litre of this space was used. Pupae were exposed in lots of 40 000 to 55 000 and if not in the chamber more than 3–4 min. no anoxia effects were apparent. Irradiation was at 9 000 r 24–48 h before adult emergence and was done in Vienna. After irradiation the pupae were placed in shallow trays with air-spaces between them to reduce to a minimum temperature rises associated with large masses of developing pupae. In this condition they were flown to Italy and samples were tested for emergence and male sterility. Some poor emergences resulted but the sterility was over 95% for the most part.

The actual field experiment on Procida is described by Murtas *et al.* (1970). The island is 3·7 square kilometres in area, and grows citrus, grapes, peaches, medlars, figs and apricots. Two-day-old fasting adults which had emerged from the irradiated pupae and had been held in paper bags of 5 000 each, were released from points on the ground 50–100 metres from each other. The adults were marked with fluorescent dyes previously applied to the pupae. 16·29 million flies were released over a period of 90 days, commencing in mid-May 1969. During this period the ratio of sterile to normal individuals ranged from 1:1 at the height of the season of abundance of the wild population (in August) to 3 776:1 two weeks after releases started. Though complete eradication was not achieved, there was good crop protection though there was some mechanical damage due to punctures by the ovipositors of sterile females. The persistence of some flies in the following season was attributed to their survival in sour orange crops over the winter.

A similar field trial by Spanish workers was made on the Canary Island of Tenerife starting in 1966, culminating in 1968 with the release of 25 million flies irradiated in the pupal stage in Madrid at 9 000 r. Fruit infestation in 1968 was only 15% as compared with the usual 70%. Later in 1969 32 million sterile flies were released over the 25 hectares of a plantation of citrus, apricot and peach in the province of Murcia in Spain between March and August, and only a 1% infestation of the fruit was recorded as compared with a 60% infestation in a control area. Again there was some evidence of damage from the sterile "stings" of released females (Mellado, 1971).

A more ambitious trial was made in Nicaragua in 1968 and 1969 involving the release from the air of 1 119 410 210 sterile flies over an area of 48 km² in the province of Carazo. The area was isolated by bait spraying a barrier-zone around it of 2 km width every two weeks with a 9:1 mixture of protein hydrolysate and malathion. Eleven million flies were released from each of four flights made each week within the barrier. After two weeks the ratio of sterile to wild insects was 116:1 and for most of the nine month release period it was in excess of 433:1. At the end of the trial pupal recovery rates per pound of fruit averaged 90·5% less than in a control area (Rhode, 1970).

The potentialities of the method for other fruit flies, e.g. Mexican fruit fly (*Anastrepha ludens*), Caribbean fruit fly (*A. suspensa*), the cherry fruit fly (*Rhagoletis cerasi*), the guava fruit fly (*D. zonatus*) and the olive fly (*D. oleae*) are now being investigated (see Sterile–Male Technique for Control of Fruit Flies, International Atomic Energy Agency Publication: STI/PUB/267, Vienna, 1970, and Sterility Principle for Insect Control or Eradication, International Atomic Energy Agency Publication: STI/PUB/265, Vienna, 1971).

Releases of irradiated Queensland fruit flies (*D. tryoni*) into isolated overwintering populations near Sydney were carried out by Monro (1966), and resulted in very rapid (two days) declines in wild populations. These were attributed to a population flushing effect rather than a sterilizing effect. The sterile insects at a ratio of 3:1 to 5:1 forced the resident populations to depart either through depletion of food resources or more likely through population pressure effects leading to actual aggression.

B. The Onion Fly

Among the other Diptera of agricultural importance irradiation sterility has been investigated in the onion fly (*Hylemya antiqua*). Ticheler (1971) has produced a mass-rearing technique enabling the production by one man of 15 000 pupae per week. Noordink (1971) has shown that pupae irradiated 1–3 days before eclosion at 3 000 r produce fully sterile females and males which in fact live significantly longer than untreated flies under laboratory conditions. The sterilized males were highly competitive under cage conditions.

C. The Cotton Boll Weevil

Some attempt to control the cotton boll weevil (*Anthonomus grandis*) by the release of irradiated insects has been made by Davich (1969). The weevils were exposed to 9 600 r and released in lots of 2 500 per hectare per week for eight weeks in two small fields near Presidio in Texas. Of 75 eggs collected in the release area during the period, only 21 hatched, and these were mostly from a very small part of one field.

D. The Cockchafer

Control and even eradication of the cockchafer (*Melolontha vulgaris*) was achieved in a 30-hectare area of grassland in north-western Switzerland by Horber (1963). Males collected from this and a neighbouring area were sterilized at 3 325 r and released on two occasions one in 1959 and the other in 1962. In the earlier trial 3 109 sterile males were released and the ratio of sterile to normal males estimated as 1:1. The white grub population dropped to about one-fifth of control areas. In the 1962 trial 8 594 sterile males were released and the ratio of sterile to normal estimated as 3:1. On this occasion complete eradication was achieved and demonstrated the feasibility of the technique with an insect which could not very well be artificially mass-reared, as the breeding cycle is of the order of three years. Jermy and Nagy (1969), however, were unable to repeat the success when they collected 53 010 wild cockchafers in Hungary in 1966, irradiated them at 3 000 r and released them in a small forest. Their results were inconclusive.

E. Lepidopteran Pests

Lepidoptera differ from most other insect orders in that very high radiation doses are needed for complete sterilization and such doses usually affect competitiveness and longevity. Lower doses, while resulting in incomplete sterility in parental matings, produce higher degrees of sterility in the F_1 generation. This peculiar characteristic is attributed to the holokinetic chromosomes (chromosomes with diffuse centromeres). When these are broken by radiation treatments, parts are not lost at cell division nor are chromosome bridges formed, but all parts pass to the daughter nuclei where recombinations in the form of translocations occur. These multiple translocations are responsible for the high degree of sterility in the F_1 generation (see Chapter 7). Thus Fossati *et al.*, (1971), working with the codling moth (*L. pomonella*), found that exposures of adult males to 30 000, 40 000 and 60 000 r gave sterilities of 85·4, 93·5 and 99·8% respectively, but reduced mating ability and longevity. An exposure to 10 000 r, however, produced only 40% sterility and had little effect on subsequent mating performance and longevity. In addition, the male progeny of untreated females mated to such males showed 91% sterility. However, mortality in the progeny of matings between irradiated males and untreated females was high irrespective of the radiation dose, and these authors consider that the actual numbers surviving to transmit the high degree of inherited sterility would be too low to have much effect. They in fact favour treatment at 25 000 r giving 75% sterility and little deleterious effect.

(i) *The Codling Moth*

Proverbs (1971), who has been carrying out field trials with radio-sterilized codling moths since 1962, at first favoured the use of fully sterilized individuals on the principle that partial sterilization would result in some fruit damage by the immediate progeny from sterile insect-wild insect matings. In his early experiments he used small abandoned orchards producing apples showing 80% infestation. Though no mechanical sex-separation technique exists for this insect, he chose to release only males in the first experiment. These were exposed as pupae 2 h before eclosion to 40 000 r in air, and the adults marked with fluorescent dust before release. The first release was made when the trees were in the spring pink-bud stage, and three releases per week were maintained until the third week in September. Releases were related to the rate of wild emergence where possible. After the first year, 957 diapausing larvae were recovered as compared with 400 in the previous autumn, and the poor results were attributed to the low ratio of sterile to fertile males of 8:1. A 20:1 ratio was intended. In the second year, however, this ratio was achieved, and only 43 larvae considered capable of overwintering were recovered. In the third year out of 2 000 males trapped, only one was a wild one, and only 6 damaged apples were found.

In a second trial mixed sexes exposed as adults to 50 000 r in carbon dioxide (equivalent to 40 000 r in air) were released together in a similar manner to that adopted in the first experiment. In the first year overwintering larvae were reduced from an estimated 5 000 to 119, and in the second year to 55. However, he estimates releases totalled 478 000 sterile moths in the second year, and found evidence of invasion from uncontrolled areas and of assortative mating.

A third trial was carried out in part of a commercial orchard where insecticide spraying had already reduced apple damage to 0·5–1·0%. Spraying was stopped in the part where releases were made, and these were of mixed sexes sterilized in the same way as those used in the second experiment. The average ratio of sterile to wild was 280:1. No damage was evident in 90 700 apples examined at the end of the season as compared with 0·5% in the sprayed area. Of interest was the fact that irradiated insects showed a higher tolerance to the insecticide than did untreated ones.

A fourth more ambitious trial involved the use of moths mass-reared on an artificial medium, irradiated as in the second and third trials and distributed from a helicopter over 48 hectares of apple, pear, apricot and cherry. The ratio of sterile to wild was above 35:1 most of the time, and less than 0·05% of the fruit was damaged. This was a better result than was being achieved by insecticide spraying.

Proverbs (1971) has recently become convinced, however, that partial sterilization may give better results than near-complete sterilization. In field cage tests he compared moths irradiated in carbon dioxide at 50 000 r with those irradiated in a similar atmosphere to 30 000 r. The irradiated moths were added at a ratio of 15 males and females to every untreated male and female. The cage containing the insects sterilized at 50 000 r actually produced 1·7 times more F_1 adults than that containing those exposed to the lower dose, showing that superior competitiveness could compensate for a lower degree of sterility.

Butt et al., (1970) describe similar experiments to those of Proverbs to control the codling moth at Yakima in the state of Washington in 1967. Newly-emerged moths of both sexes were inactivated and sterilized by exposing them to 40 000 r at 1°C. Over a 6-month period 1 541 745 were released into an isolated 93-acre apple orchard. Most were dropped from a helicopter. Control was considered as good as that given by the use of insecticides. The average ratio of sterile to wild insects was 67:1.

(ii) *The Cabbage Looper*
Henneberry (1971) describes the results of irradiation experiments with the cabbage looper (*Trichoplusia ni*). Three-day old males need 40 000 r exposures to produce 100% sterility, while 15 000 r gives only 32–55%

sterility, yet results in 97% sterility in the F_1 generation. A comparison in large field cages of partially sterile (15 000 r) and near-fully sterile (30 000 r) males at a ratio of 9:1 with fertile females, showed that the former reduced the population by 90% and the latter by 80%. North and Holt (1971) working with the same insect found that uninseminated females exposed to 20 000 r were 93·4% sterile, while their male progeny were only 58·7% and their female progeny only 16·6% sterile. Inseminated females on the other hand were completely sterile after exposure to 15 000 r and they advocate as a practical method of control the setting up of a number of population cages, allowing mating and a first oviposition to take place and then sterilizing the remaining contents at 15 000 r, setting up further cages from the oviposition. A large-cage trial was in fact done on these lines, and a 9:1 ratio of sterile to wild resulted in a hatch of only 9% in the first generation and 53·5% in the second generation. This represents a total control of more than 90% in the two generations.

(iii) The Gipsy Moth

In Jugoslavia a small-scale attempt has been made by Maksimović (1971) to control an isolated population of the gypsy moth (*Lymantria dispar*) by the release of males irradiated as 9–12 day old pupae at 30 000 r. This dose gave full sterility. An area of 1·3 hectares on the island of Hvar was used, and an estimated ratio of only 0·5 sterile to one normal male achieved with a release of 289 males. Notwithstanding, 22% of recovered egg-masses showed hatches of less than 30%.

(iv) The Corn Earworm

A field release of irradiated corn earworm (*Heliothis zea*) males on the island of St. Croix in the United States Virgin Islands is described by Snow *et al.*, (1971). The insects were reared in Georgia and despatched as pupae to the island. There emerging males were irradiated at 33 000 r and released from seven ground sites between April and July 1969. Although the island has a total area of 84 square miles, only three to four acres of it were taken with corn. Monthly slumps in the supply of material from Georgia due to pathogen problems in the colony caused marked fluctuations in releases, but when deliveries were normal 1 100–5 000 males were released each day, and estimated ratios of sterile to normal males during these times ranged from 20:1–50:1. Good control shown by a marked reduction or complete lack of eggs from corn plant inspections occurred during peak releases, but rapid population recoveries occurred during slump periods. The authors suggest that the lack of eggs was due not to sterility of the males as such but a locking in copula of sterile male and wild female which they were able to show as occurring when seven virgin females from the wild were caged with five sterile males. Four failed to separate after copulation.

III. Public Health Pests

A. Houseflies and Related Flies

Surprisingly few attempts have been made to control houseflies and related flies by the release of irradiated insects. Rivosecchi (1962) attempted to control the common housefly (*Musca domestica*) in a coastal area of Latina province in Italy by releasing 45 500 males sterilized at 2 000 r between April and July. Sterile females began to show themselves one month after releases started, and in the following two months the female flies were markedly reduced. However, from August onwards there was an increase in fertile flies again, and it was concluded that the area was insufficiently isolated for eradication to be achieved and that the sterile males were not competitive enough. MacLeod and Donnelly (1961) tried without success to eradicate the blowfly (*Lucilia sericata*) from a small island of some five square kilometres off the north-east coast of England (Holy Island) during the years 1956 to 1958. A sterilization dose of 6 000–7 000 r was used, but this reduced longevity by 40% and hence competitiveness. Donnelly (1965) estimates that normal males successfully copulate six and sometimes twelve times, while the sterile male is for the most part only capable of copulating once. In addition, this copulation does not necessarily prevent the female from subsequently accepting a normal male, nor from producing viable offspring.

B. Mosquitoes

Mosquito control by the irradiation method has been tried on a number of occasions but with little success. Three of the trials were carried out over ten years ago. The first concerned *Anopheles quadrimaculatus* in Florida, and involved the release over a period of fourteen months of 433 600 males irradiated as less-than-24-h-old pupae at a dose of 12 000 r. Sex separation was performed by hand after adult emergence. No reduction in total numbers of mosquitoes nor increase in sterility of wild females resulted, and it was concluded that the laboratory strain used for mass-production was no longer competitive, because it had been laboratory-bred for some 200 generations over the previous 25 years (Weidhaas et al., 1962). An attempt to control *Culex pipiens fatigans* in India by the release over a period of 35 days of 24 000 males irradiated at 7 000 r resulted in only a 6% reduction in the hatching of egg rafts, and was discontinued because of objections from the people in the village where the trial was made (Krishnamurthy *et al.*, 1962). Finally a field release of 4 777 000 male *Aedes aegypti* exposed as young pupae to 11 000–18 000 r was made near the town of Pensacola in Florida. In this case the male pupae could be separated mechanically from the female ones because of a marked difference in size. Releases extended over a period of 43 weeks but produced little effect on the wild population (Morlan *et al.*, 1962). Failure has been attributed to too high an irradiation dosage affecting male fitness and the inadequate dispersal of release points. Too little account was taken of the short flight range of this species. Weidhaas and Schmidt (1963) subsequently showed that males derived from pupae sterilized at 10 000 r did not compete at all with normal males in laboratory cages, and those exposed to 8 000 r only poorly. When not in competition with normal males these irradiated males mated just as readily and lived as long.

More recently two additional attempts have been made to control *C. p. fatigans* in India. Laboratory studies by Sharma *et al.* (1972)[1] showed that 6 000 r applied to male pupae gave 99% sterility, and that if applied to pupae more than 24 h old produced highly competitive males, though not as competitive as young adults so exposed. Pupal exposure resulted in some reduction in longevity though this was not marked in the important first week or 10 days of life. No recovery in fertility could be found two weeks after irradiation. Females exposed as pupae to 5 000 r and above failed to oviposit. Thus the slight inaccuracy of the pupal sexing technique would result in the release of a few females, but these would be completely sterile. The release of marked males emerging from irradiated pupae showed them to disperse as quickly and as widely as untreated mosquitoes (Rajagopalan *et al.*, 1973).

For the first field release in March, 1971, a small village, Sultanpur, just outside Delhi was chosen. This was composed of some 200 houses and had a population of some 1 700 people. At the time of the release the daily emergence of the wild population was of the order of 24 000 to 30 000 individuals of both sexes each day. Male pupae irradiated at 6 000 r when 24–36 h old were released from clay pots or from actual breeding places over an eight-day period at the rate of nearly 9 000 per day. A 10% mortality in treated pupae was estimated, and the emerging males were reckoned to be only 50% competitive. Thus the functional ratio of sterile to normal males was calculated to be more like 1:3 than the intended 1:1. However, the 25% sterility expected from a 1:3 ratio was achieved by the fourteenth day after the start of releases and a significant sterility persisted for at least another sixteen days (Patterson and Sharma, 1972).[2]

[1] Sharma, V. P., Patterson, R. S., Seetharam, P. L. and LaBrecque, G. C. World Health Organization mimeographed document WHO/VBC/72. 345.
[2] Patterson, R. S. and Sharma, V. P. World Health Organization mimeographed document WHO/VBC/72.339.

A second release was made in September of the same year in the nearby village of Pochanpur.[1] In this case adult males 1–2 days old were irradiated at 6 000 r, and 61 400 to 88 800 were released every two days for 4 weeks into a population whose original size was estimated to be 14 860 individuals with some 2 000 males emerging each day. In the third and fourth weeks after releases started the sterility in recovered egg rafts was over 60%.

C. Reduviid Bugs

Attention is now being paid to the possibility of irradiating *Rhodnius prolixus*, one of the vectors of *Trypanosoma cruzi*, the causative agent of Chagas disease in South America. Gomez-Núñez (1971) found that males of this species exposed to 10 000 r after feeding and emergence to adult are 92% sterile and almost equally competitive with normal males, though their length of life is shorter. 5 000 r produced only 70% sterility and both exposures led to abnormal mortality in the F_1 nymphs derived from matings between irradiated males and untreated females. Baldwin and Chant (1971) found that newly-produced males of this same species were much less affected by exposure to gamma rays in an atmosphere of nitrogen than they were in air. All survived 10 000 and 15 000 r for fifteen weeks after treatment and readily mated. In air the same treatments killed the insects or affected their mating efficiency within the first seven weeks.

IV. Radiation Resistance

Evidence of the ability to develop resistance to the lethal effects of irradiation has been found in *Ae. aegypti* (Terzian and Stahler, 1966; Stahler, 1971). A laboratory colony of some 20 years standing was selected by exposing 2–4-weeks-old eggs to 2 000 r in the first 10 generations, and 2 500 r for a further 80 generations. Initially this caused more than 90% aquatic stage mortality. In each generation the percentage yield of adults was compared from irradiated and non-irradiated eggs of the selected strain and an unselected control. In the selected strain the percentage yield from irradiated eggs rose from less than 10% to nearly 40% in 16 generations as compared with a more than 90% yield from unirradiated eggs of both this and the control strains. Throughout the 90 generations the control eggs only yielded an average of 7% adults after irradiation. The resistance remained at much the same level between F_{16} and F_{50} but from F_{51} to F_{64} there was a further increase to around 60% of irradiated eggs producing adults. This yield was maintained up to the F_{90} though the yield from

---

[1] Information from monthly reports of the WHO/ICMR (Indian Council of Medical Research) Research Unit on the Genetic Control of Mosquitoes, Dehli.

unirradiated eggs of the selected strain declined from around 94%–82%. There was no change in the yield of unirradiated eggs of the control throughout the 90 generations. The selected strain was shorter lived than the unselected one (Terzian and Stahler, 1966) and though blood-feeding increased the longevity it never equalled that of the normal non-blood-fed one (Stahler and Terzian, 1968). In addition, the frequency of oviposition was reduced in the selected strain as well as the susceptibility to chicken malaria (Stahler and Terzian, 1966). The effect of selection on the production of dominant lethals was not studied.

4. Chemosterilants

Any chemical compound affecting fertility may be called a chemosterilant. A large number of such chemicals exist and act in different ways. Antimetabolites for example inhibit the utilization of essential metabolites such as folic acid, glutamine, purine and pyrimidine. Their greatest effect is on the female and usually results in infecundity. Organo-metals, for example, organo-tin compounds, developed first as fungicides and molluscicides have shown sterilizing as well as anti-feeding properties when given to some insects, though none have yet been seriously considered for practical use in insect control. Juvenile hormone analogues are also known to affect insect fertility, though they are better known for their effects on moulting. Masner *et al.* (1968) showed that the female lime bug (*Pyrrhocorris apterus*) kept with males treated with $100\,\mu g$ of DMF (dihydrochloride of methyl farnesoate) laid only sterile eggs, and that even 1-μg-treated males reduced egg viability by 50%. Patterson (1971) tested three juvenile hormone mimics (CJH 2, FME and FAE) topically applied to inseminated, blood-fed *Ae. aegypti*, and found that significant sterilization could result, if application was made half way through the gonotrophic cycle, but not immediately after the blood-meal, nor when the female was fully gravid. The monogamy-inducing hormone of the accessory gland of many male insects has already been mentioned. Sublethal concentrations of insecticides can also affect fertility as can a whole host of other chemicals, such as colchicine, reserpine, griseofulvin, sterculic acid, ethylene glycol and a number of urea and thiurea derivatives. However, what we are concerned with from the point of view of the genetic control of insects are those chemicals which are capable of producing dominant lethal mutations. As LaChance (1967) points out "all chemicals that induce dominant lethal mutations are chemosterilants but not all chemosterilants can induce dominant lethal mutations—or mutations of any kind for that matter."

Most of the dominant lethal-producing chemicals are alkylating agents, that is to say, substances producing carbonium ions ($-CH_2$) which in living tissues combine with some of the nucleic acids and proteins of the cell. This in some way upsets the genetic code and leads to point mutations and chromosome breakage (Fahmy and Fahmy, 1964).

The fact that chemicals could produce genetic changes in the same way as radiation was not discovered until the early years of the Second World War when Auerbach and Robson (1947a, b) found that mustard-gas could produce mutations in *Drosophila* as well as affect fertility. Independently, Rapoport (1947) discovered that ethyleneimine was also mutagenic, and it is from analogues of this compound—aziridines—that most of the modern insect chemosterilants are derived, for example, tepa, metepa, thiotepa, tretamine and apholate. However, the potentiality of alkylating agents as insect sterilants was not recognized for at least another ten years. Rather, some of them were to be used as palliative treatments for various forms of cancer as well as strengtheners of synthetic fibres and for the flame-proofing of natural fibres.

As with radiation, it is important to distinguish between the sensitivity of germ cells to injury and to the induction of dominant lethals. It will be remembered that in general the earlier stages of development of the germ cells are more susceptible to radiation injury than later ones. This is not necessarily the case with alkylating chemicals. Tri (ethyleneimino) triazine (TEM) is only half as injurious to spermatogonia as to spermatozoa in *Drosophila*, while phenylalinine mustard is four times as injurious to spermatogonia (Fahmy and Fahmy, 1964). This means that insects exposed to TEM are more likely to recover fertility than those exposed to the amino-acid mustard. In terms of dominant lethals both radiation and chemicals are more likely to produce them in primary spermatocytes and in oocytes in anaphase I than in other stages of development.

In general the final effects of radiation and dominant-lethal-inducing chemicals at minimum sterilizing doses are very similar. The only difference LaChance and Riemann (1964) were able to demonstrate with the screw-worm, when irradiation and tretamine treatments producing nearly 100% dominant lethals were given to adult females with oocytes in early prophase or early anaphase, and males containing mature spermatozoa, was in the timing of interruption of embryogenesis. In irradiated insects most of the damage occurred during the first and second meiotic divisions of the egg (which, it may be remembered, occur after oviposition in insects). In tretamine-treated insects no such damage was evident until the early cleavage divisions following meiosis.

As with radiation treatments alkylating agents can produce aspermia, sperm inactivation and somatic cell damage. They can also affect longevity and mating competitiveness, though the weight of evidence indicates that most chemicals are less harmful in these respects than are radiation treatments. In fact as we shall see there is evidence that some chemosterilized insects may be more sexually active than untreated ones.

Chemosterilants offer a number of advantages over irradiation sterilization. They are relatively cheap and do not require expensive apparatus to apply them, where the sterilization of mass-reared insects is involved, at any rate. They can be applied in a number of different ways—orally, topically, by injection, by spraying, by dipping, by fumigation or by exposure of the insect to a treated surface. Many of these methods are impractical for large-scale work and all must involve considerable variation in the actual concentration of chemical reaching the sites of action. Thus in all probability working concentrations are over-concentrations for most of the individuals exposed at any one time, and even with these over-concentrations more escapes from sterilization must be expected than would be the case from the more uniform method of application of radiation treatment. This then must be taken as a slight disadvantage.

One enormous potential advantage of sterilizing chemicals is the possibility of their use in the field to sterilize wild populations. This would obviate the need of a mass-rearing facility but requires the controlled use of chemicals dangerous to man and his environment. The solution to the problem lies in the combination of sterilant with attractant. This could be a chemical attractant such as a sex pheromone or some form of food or a physical attractant based on light or sound for example. Such attractant traps present additional costs of course. Unfortunately, really effective sterilant-attractant combinations have yet to be found for most insect pests and, for the present, mass-rearing and sterilization under strictly controlled laboratory conditions are the normal procedures.

The great disadvantage of alkylating agents is that their sterilizing and mutagenic effects extend to higher animals including man, and that some of them at least are carcinogenic and even phytotoxic (see Campion, 1972). Fahmy and Fahmy (1964) estimate that 0.1 mg per kg body weight of TEM would be the equivalent of 35 r of ionizing radiation. This is the "doubling radiation dose", the dose needed to double the spontaneous human mutation rate or to add to the human gene pool one additional mutation per five individuals per reproductive generation (30 years). For a weaker agent like Myleran (a sulphonic ester) 20 mg/kg would be required to produce the same effect. So concerned are the toxicologists with the dangers of some of these chemicals that investigations have been made of their persistence in treated insects on the slight chance that those having subsequent contact with man might contaminate him. Dame and Schmidt (1964) for example using P^{32}-labelled metepa recorded an uptake of $2.5\,\mu g$ and $7\,\mu g$ per female for *Aedes aegypti* and *Anopheles quadrimaculatus*, respectively, when exposed for 4 h to a deposit of 10 mg of metepa per square foot on glass, but most of this was lost within one day. Plapp *et al.* (1962), using metepa-treated *Culex tarsalis* and *Musca domestica*, found that the compound was completely degraded in the former insect in 24 h, and in the latter in 48 h, with indications that in the mosquito persistence in its sterilizing form was only for 6–8 h after treatment. More recently, LaBrecque *et al.* (1972b) have investigated the persistence of thiotepa and its metabolite tepa in *C. p. fatigans* treated as pupae by immersion in an 0·6% aqueous solution of thiotepa for three hours. Estimations were made, using gas chromatography, of residues in pupae and emerging adults at various intervals after immersion and subsequent rinsing in clean water. An average of 62·2 nanograms (ng) of thiotepa per pupa was found immediately after treatment, falling to 12·0 ng 24 h later. No tepa was found in the pupae, but 2·0 ng per insect was recovered in adults up to 6 h old. These also showed 10·2 ng of thiotepa per insect. Adults 24 h old or older showed no observable residues of either compound. Thus the release of 24 h old adults would be completely safe, and it would take 80 000 pupae 24 h old or 100 000 newly-emerged adults to accumulate 1 mg of thiotepa. Campion and Lewis (1971) on the other hand showed that topically-applied tepa was lost quite slowly from the red bollworm (*Diparopsis castanea*)—50% of 10 µg in 72·3 hours. This compared with 19·3 h when the same amount was injected.

Attempts to find safer chemosterilants have resulted in the discovery of non-alkylating relatives of tepa and tretamine (melamine), viz. hexamethylphosphoramide (hempa) and hexamethylmelamine. Hempa is the most tested and seems effective against *M. domestica*, *Ae. aegypti* and the *C. pipiens* complex, but virtually inactive against the boll weevil (*Anthonomus grandis*), Japanese beetle (*Popillia japonica*) and the milkweed bug (*Oncopelta fasciatus*). Only weak activity is shown with the screw-worm, codling moth, red bollworm and many of the fruit flies (see Campion, 1972).

In the remainder of this chapter examples will be given of the treatment of various insect pests with various dominant-lethal-producing chemosterilants with emphasis on those procedures developed to the point of field testing. The whole treatment of chemosterilants is the subject of a book edited by LaBrecque and Smith (1968), and a more recent comprehensive review has been produced by Campion (1972). This author lists 125 insect species as having been treated with chemosterilants of one kind or another, and his article has a bibliography of more than 600 titles.

I. Public Health Pests

A. Houseflies and Related Flies

Most of the early work, and especially the field application, with chemosterilants was done against houseflies. LaBrecque *et al.*, (1962a) showed that male flies sterilized on a diet of 1% apholate seemed to be more successful in mating with females than untreated ones. With a 1:1 ratio of sterilized to normal flies the reduction in fertility of females was 61% rather than the expected 50%. Further, Smith *et al.* (1964) showed that these apholate-sterilized males could mate repeatedly with virgin females for 28 days after sterilization, that they still had motile spermatozoa in their testes and could still produce 100% sterility in females mated by them. In the first field experiment a commercial bait (of cornmeal, sugar, powdered milk and egg) containing 0·5% tepa was put on a semi-isolated, one-and-a-half-acre, refuse dump on Bahia Honda Key in the Florida Keys. The dump was treated weekly for nine weeks (with the exception of the second week), and density measurements made by counting the numbers of flies landing in one minute on an 18 × 18 in, grid in ten scattered localities showing the highest densities. In four weeks this number dropped from an average of 47/grid to 0. Over the same period the percentage hatch of eggs laid by captured females fell to 1%. That isolation was not complete, however, was shown by an increase in egg viability up to 50% and an increase in fly abundance to three per grid four weeks after treatment of the dump ceased (LaBrecque *et al.* 1962b). Gouck *et al.* (1963) repeated the experiment on Puie Island, Florida, using a similar bait, but containing 0·75% apholate. Forty lb. of bait were applied each week for seven weeks to a two-acre refuse dump, and then 15 lb each day for five days of each of the following five weeks. The grid count of flies was reduced from 68 to near zero, and the egg-hatch from 81% to 1·2%. The male fertility averaged only 22% in the last five weeks, but again eradication was not achieved because the area was insufficiently isolated. LaBrecque *et al.* (1963) and LaBrecque and Meifert (1966) transferred their attention from refuse dumps to poultry houses, and applied 0·5% and 1·0% metepa, 2·5% hempa and 1% apholate in various baits to the droppings under the cages. Twice weekly applications gave a high degree of sterility and control, and in one isolated area complete elimination was claimed with only three flies being seen during two and a half months after treatments ceased. Later, Meifert and LaBrecque (1971) produced 98% sterility in flies originating from poultry droppings treated with a mixture of dimethoate (insecticide) and 1% hempa in sugar solution, and more than 90% reduction in abundance in 20 weeks. Dimethoate by itself at 1 g/m² gave adequate control if applied twice a week, but resistance was likely after prolonged use.

Further experiments to control houseflies were made by Meifert *et al.* (1967a) on two West Indies islands by the twice-weekly application of chemosterilant baits to privies. On Grand Turk 1% metepa liquid bait induced a sterility in excess of 80%, and over one and a half years a more than 90% reduction in the fly abundance. On Mayaguana a 1% apholate liquid bait was less effective and produced 60–80% sterility over three months and 50–80% less flies. These results compared with a 50–90% reduction in a third island, San Salvador, where the privies were treated with the insecticide trichlorfon.

In contrast Hansens (1965) and Hansens and Granett (1965) working with large-cage populations of houseflies, only reduced one population by 75% in three months using 2% apholate in a sugar bait while the insecticide trichlorfon almost eliminated another population in one week. More recently Pausch (1971, 1972) after experimenting with different baits and establishing that hempa, metepa and tepa were effective sterilants in laboratory trials, succeeded in completely eradicating an enclosed population of houseflies in a barn by the use of 3% metepa in a sugar syrup bait. A sterility of more than 80% was achieved in four weeks in this trial.

A unique way of applying a chemosterilant to the housefly was devised by Morgan (1967). This involved the attachment of a small chamois pad impregnated with metepa to the abdomen of the female, so that males subsequently copulating with such "booby-trapped" females would sterilize themselves. Whitten and Norris (1967) suggested a similar method using insecticide instead of chemosterilant for the control of *Lucilia cuprina*. They showed that a female dieldrin-resistant *L. cuprina* could be topically treated on the thorax with $0.5\,\mu l$ of 2% dieldrin without lethal effect. When caged with susceptible males such females could kill as many as 100 males. Thus, they suggest, sterilized insecticide-resistant females, "booby-trapped" and released into wild populations could introduce not only sterility, but also death to susceptible insects.

In Italy several attempts have been made to control houseflies with chemosterilants. Sacca and Stella (1964) sprayed the garbage dumps of small towns with $0.0625–0.2\%$ concentrations of tepa in syrup with 1% malt extract as additional attractant. The sterility status of both sexes was determined and females showed a higher degree, presumably because they contacted the dumps more often (for oviposition for example) in addition to being mated by sterile males. A total application of 64 g of tepa in 94 l of syrup given as once weekly sprayings in the first month, and twice weekly in the second, prevented the spring build-up in fly population. Sacca *et al.* (1966a) also claimed good control when 1.25 and 3.75% concentrations of hempa were used on such dumps.

On Vulcano, one of the Lipari Islands off the north coast of Sicily, temporary eradication of *M. domestica* has been achieved by the combined use of insecticide in houses and animal shelters (plastic strips impregnated with the organophosphate, dichlorvos) and the release of tepa-sterilized males. The fly population was estimated to have been reduced from 500 000 to 20 000 by the insecticide before the weekly release of 20 000 sterilized males started. These were fed during the 24–36 h prior to release on a

0·05% solution of tepa in sugar syrup. Weekly releases of 20 000 took place from July to October 1967 and then insecticide treatment stopped. At that time the fly population stood at some 40 000 individuals of which 28 000 were released sterile males. From November 1967 to June 1968 a further 989 000 sterilized males were released and from December eighth no females could be found. However, evidence of re-invasion was found in the following April (Magaudda et al. 1969).

Good control of M. domestica vicina is reported by Japanese workers (Matsuzawa and Fuji'i, 1968) from a field trial of hempa in milk-bait on Sei Island in the Seto Inland Sea between August and October 1967. The expected autumn rise in numbers did not appear.

A continuous search for new, better and less hazardous chemosterilants has been going on since 1958, especially in the United States. One in particular may be mentioned, N,N[1] tetramethylenebis (1-aziridine carboxamide) (TMAC), which holds promise for both screw-worm and housefly control. The screw-worm seemed to be one of the few exceptions to the rule that insect fitness was more affected by radiation treatments than by chemosterilants. Most of the common chemosterilants were less efficient against this species than irradiation. Crystal (1965) found TMAC to be particularly good, however, and even found it to cause an increase in mating competitiveness of males. Sacca et al., (1971)[1] found that male houseflies fed on 0·001% solution of the chemosterilant in 20% sugar solution were 95% sterilized, and estimate it to be 159 times more effective than tepa and 370 times more effective than apholate. Females, however, are some 100 times less sensitive to TMAC than males. The males retain a sterility of 85–90% one month after treatment, and are more sexually active than untreated males though not as markedly so as screw-worm males. A field experiment is being carried out on the Italian mainland with this compound, but has not yet been reported upon. No information on human toxicity problems is given. It might be mentioned that Meifert et al. (1967b) successfully "booby-trapped" female M. domesticus with this compound causing sterility in males subsequently caged with them.

Other methods than the feeding of chemosterilants to houseflies have been tried in attempts to improve the proportion of flies receiving sterilizing dosages, presumably. Fye et al., (1968) had flies migrating through expanded polystyrene foam strands treated with 5% tepa. Good competitiveness of males resulted even when they had to go through 18 cm of such strands, and no recovery in fertility occurred during the four ensuing weeks. Fye and LaBrecque (1971) had emerging flies walking over chemosterilant-treated surfaces, and showed good sterilization when tepa and metepa were used but not with hempa.

[1] Sacca, G., Mastrilli, M. L. and Pierdominici, E. (1971). World Health Organization mimeographed document. WHO/VBC/71.329.

Sacca *et al.*, (1969) found evidence of a synergistic effect between tepa and hempa when used against house flies. Mixtures given orally increased the activity of tepa twenty times and hempa five times.

Some evidence of resistance of houseflies to chemosterilants exists. Abasa and Hansen (1969) selected a laboratory population with apholate for 35 generations, beginning with a concentration of 0·05% in sugar bait. Resistance first became apparent in the fifteenth generation, and was ten-fold by the twentieth and twenty-six-fold by the thirty-fifth. Then the flies could be fed on a 1·9% concentration without apparent effect. Egg fertility observations showed the selected population to be about five times more resistant than the parent with a low level of cross-resistance to metepa. Sacca *et al.* (1966b), on the other hand, selected through 22 generations with metepa and hempa. No resistance to the latter was apparent and only a slight tolerance to the former. This tolerance was lost when selection stopped. In addition, Morgan *et al.*, (1967) maintained a colony on a diet containing 0·01% apholate for 65 generations. The sterility of females increased at first from 6% in generations 1–5 to 69% in generations 26–30, and then decreased to 14–22% in generations 51–65. These authors did not consider this to be real resistance. Sacca *et al.* (1965) found some evidence that metepa was repellent to *M. domestica*.

LaBrecque *et al.* (1972a) have released stable flies (*Stomoxys calcitrans*) sterilized by immersing four-to-five-day-old pupae in a 5% aqueous solution of methiotepa for one hour in Florida at a ratio of 1:1 with the wild population. A marked reduction in the field population resulted.

B. Tsetse Flies

Field trials with chemosterilized tsetse flies (*G. morsitans*) have been carried out in Rhodesia. Prior laboratory trials had indicated that tepa and metepa were effective sterilants while apholate and hempa were not (Dame, 1968). Contact exposures or sprays delivered in a wind tunnel resulted in permanent sterility without effect on male longevity or mating competitiveness using tepa and metepa, and both were also effective when used as pupal dips. Sterilized flies proved effective transmitters of *Trypanosoma congolense* though less efficient than untreated ones. Sterilization after infective feeds reduced transmission more than if done before such feeds. Three field releases were made on islands in Lake Kariba (Dame and Schmidt, 1970). In the first in 1967, males emerged from field-collected pupae were exposed to 10 mg/m² tepa for 30–60 min. and then released. Prior to release the wild population was estimated to be of the order of 4 000 per square mile and two rounds with the insecticide BHC had reduced it to about 2 000. Releases were maintained for 180 days but ratios of sterile to fertile males determined from recaptures were never greater than 12:100. The survival of the released males was only 17% of that of the wild males. No control resulted. A second experiment was made on another island with an estimated population of 600–1 200 *G. morsitans* per square mile. 26 000 males sterilized as before were released over a period of one and a half years, and the last female was seen three months before the last release. However, some of the population decline was attributable to a decrease in host animals. The third trial was carried out in 1968 on the same island as the first trial at a time when the tsetse population was estimated at 3 000 per square mile but had been reduced to about 300 by two aerial applications of dieldrin. Sterilization was achieved by immersing puparia in 5% tepa and the sterilized insects were allowed to emerge "on site". 98% control was achieved in nine months, though never complete eradication.

C. Mosquitoes

(i) *Aedes aegypti*

Like the house fly, mosquitoes have been the insects for tests in many studies and field experiments with chemosterilants. *Ae. aegypti* and *C. p. fatigans* are the two principal species involved. Bertram (1963) carried out a series of laboratory observations on the former species using the sulphur analogue of tepa, thiotepa [tris (1-aziridinyl) phosphine sulphide]. Working first with blood-fed, inseminated female *Ae. aegypti* up to one week old, Bertram found a 3-h exposure to a 20 mg/sq. ft. deposit was needed to give 100% sterility in eggs subsequently laid; 3 h on 10 mg/sq. ft. gave only 77% sterility. Three hours on 40 mg/sq. ft. did not affect survival for two weeks, but in the third week the mortality was higher than in controls. Males proved more sensitive both to handling and to the chemosterilant. Three hours on 40 mg/sq. ft. caused 94% mortality within one day. Survival after the first day was good for two weeks, however, after a 3-h exposure to 20 mg/sq. ft., and most of the males were still fully sterile at the end of this period. Active sperm were recovered in testes 32 days after the exposure, however, and matings then showed some recovery in fertility. Virgin females mated to sterilized males remained sterile up to 30 days, but virgin blood-fed females exposed to 20 mg/sq. ft. for 3 h and a day later mated to normal males gave a 40% egg hatch. This observation indicates that most of the sterilization of already-inseminated females is due mainly to the effect on the spermatozoa and only partly to the effect on the ova. In other words the ova are less susceptible to the sterilant than the spermatozoa. However, marked progressive reductions in clutch sizes were apparent in successive gonotrophic cycles in sterilized females. Bertram (1964) records the average egg-batch size as 94·4 in the first gonotrophic cycle in untreated females, 73·7 in the second and 57·1 in the third. Equivalent figures in females exposed to 40 mg/sq. ft. for 3 h were 72·2, 1·1, and no eggs were laid after this tiny second oviposition. Females exposed within 24 h of emergence for 3 h to 20 mg/sq. ft., never developed mature eggs, but 24-h-old females developed their eggs normally. Bertram concludes from this that though the oocyte itself may not be directly damaged, the surrounding follicle cells may be, and these will have a greater effect in subsequent cycles. In the light of what has been said in the section on sterilization in Chapter 2, the explanation might be in a difference in the degree of development of the nurse cells.

Bertram *et al.* (1964) studied the effect of exposure to thiotepa of the *Ae. aegypti* infected with the chicken malarial parasite *Plasmodium gallinaceum*, and were only able to reduce transmission of the parasite by 25% with exposures of 3 h to 20 mg/sq. ft., just after the infective feed. Three hour exposures to 40 mg/sq. ft. just after the feed produced a high mortality in the mosquitoes, as did a 2-h exposure to the same deposit 24 h after the blood-meal. Two hours on 40 mg/sq. ft. immediately after feeding produced no decrease in transmission of the malaria parasite. Those infections delivered to chickens by chemosterilized females did not seem to differ in pathogenicity in any way from those given by untreated females. The authors conclude that the vector is more easily sterilized than the parasite. These findings contrast with those of Altman (1963) (who it might be said used different strains of both parasite and mosquito), who showed an 85% reduction in transmission by mosquitoes exposed to 10 mg tepa/sq. ft., on glass for up to 9 h before or after an infecting meal.

Bertram (1964) also showed that the growth of the filarial worm (*Brugia patei*) in thiotepa-sterilized *Ae. togoi* was so markedly affected that transmission would be unlikely by fully sterilized females.

A much more practical way of achieving the sterilization of mosquitoes on a large scale is by the immersion of pupae in solutions of chemosterilant. This was first done by White (1966) using *Ae. aegypti* and thiotepa. Twenty-eight hours in 100 parts/10^6 gave complete sterility in emerging males, but only 51% sterility in females. Twenty-four hours at 500 parts/10^6 gave better results with females, and even resulted in some inhibition of ovarian development. Twenty-four hours in 1 000 parts/10^6 resulted in complete sterility and marked infecundity. Providing that pupae were more than one hour old and less than one hour away from eclosion, age had little effect on subsequent sterility. Males emerging from pupae exposed to 100 and 200 parts/10^6 for 24 h were highly competitive, and no loss of sterility was evident up to 16 days. In fact, evidence of increased activity in the chemosterilized males was later demonstrated (White, 1970).

Apholate and hempa have also been tried on *Ae. aegypti*. Weidhaas and Schmidt (1963) allowed males to feed on 1% apholate in a 20% aqueous honey solution for the first three or four days after their emergence, and found them highly competitive. Judson (1967) found the same chemosterilant to inhibit ovarian development in females and that such females showed an increased biting activity. He therefore suggests that the release of such females might lead to a temporary increase in disease transmission before a population decline became evident. Powell and Craig (1970) found evidence of an inhibition of the development of the accessory glands in male mosquitoes emerging from larvae reared in 20 parts/10^6 apholate. They found the volume of the accessory gland in such males was only half that of untreated males, and they were not as efficient in rendering the females monogamous, though their actual mating ability was not affected; apparently they could inseminate nearly as many females as untreated males.

Grover and Pillai (1970a) studied the effect of hempa on *Ae. aegypti*, and showed that sterilized males were fully competitive, but there were indications that they depleted their spermatozoa faster than normal ones and that there was some evidence of recovery of fertility after multiple matings.

There is some evidence of resistance to apholate and metepa in *Ae. aegypti*. Hazard *et al.* (1964) selected larvae for eleven generations at concentrations of apholate which sterilized 50–90% of the eggs of emerging adults, and produced a four to five-fold increase in tolerance, while Patterson *et al.* (1967), record that after 43 generations this resistance was 20-fold with three-to-four-fold cross-resistance to metepa. Klassen and Matsamura (1966) found a three-fold resistance in the same species after eight generations of larval selection with metepa. With hempa, however, George and Brown (1967) were undecided whether the temporary change in tolerance they recorded was real resistance. They exposed fourth instar larvae for 24 h to 640 parts/10^6 for three generations and subsequently to 1 280 parts/10^6. The initial level of sterility was 67% from exposures to 640 parts/10^6 and 83% from exposures to 1 280 parts/10^6. This latter level declined to 68% in the F_5 generation after the exposure of F_4 larvae to the higher selection concentrations. However, this apparent resistance was lost in the F_6 generation, and it is concluded that any tolerance due to detoxification is eventually offset by the accumulation of recessive genetic defects attributable to the mutagenic effects of the chemosterilant.

(ii) *Culex pipiens fatigans*

Early experiments with *C. p. fatigans* involved the exposures of aquatic stages to aqueous solutions of chemosterilants. Mulla (1964) exposed fourth instar larvae to 10 parts/10^6 of apholate and showed 96% sterility in emerging adults while similar exposures to the same concentration of tepa and metepa only produced 16% and 5% sterility respectively. Murray and Bickley (1964) showed that the males emerging from fourth instar larvae exposed to 10–15 parts/10^6 of apholate were quite competitive with normal males. Mulla (1964) further showed that adults offered 0·01% solutions of apholate, tepa and metepa in sugar water for 24 h did not produce any offspring subsequently.

Laboratory experiments involving the inactivation of newly-emerged males at low temperature and their rotation in capsules with apholate dust showed this to be an effective sterilizing method providing the dust contained 75% or more of the chemosterilant. These males remained sterile over eleven days after treatment and were fully competitive. Patterson et al. (1968a), went on to try simulated releases of such apholate-dusted males in a large, divided, outdoor cage, each section containing its own breeding pool. Treated males were released in one half; the other served as a control. Sample dips from pools served to assess the daily mosquito emergence rate and daily releases varied from 250–1 000 males over a two-month period. Theoretical ratios of sterile to normal males were calculated to vary from 5:1 to 40:1 though the actual ratio was not determined. In the first month when the average ratio was 15:1 the average sterility of collected egg-rafts was 27%. In the second month these figures were 23:1 and 50%. There was evidence of a decreased survival of treated males due as much to the method of application as to the chemosterilant itself probably. Even so the results might seem disappointing as the expected sterility from such ratios is much higher. However, these must be over-estimates of actual ratios as they do not allow for the number of males present at the start of the experiment and the superior survival of untreated males. Patterson and Lofgren (1968) repeated these trials a year later at first allowing males to emerge through tepa-treated polystyrene strands but later having to revert to apholate dusting through lack of tepa. Nevertheless, better results ensued though not eradication. Apholate and tepa laboratory testing was continued by Pillai and Grover (1969) and Grover and Pillai (1970b) who showed that sterilized males were more competitive than normal ones but that apholate-sterilized males showed a greater tendency to recover fertility.

Patterson et al., (1971) transferred their attention to thiotepa as a pupal treatment and found that a four-hour immersion in 0·7% aqueous solution gave good sterilization of males. Tepa and metepa did not give good results with pupae, but males emerging through polystyrene strands or polyvinyl tubes dipped in 5% tepa in methanol were effectively sterilized. Further trials with the two-part outdoor cage showed tepa-treated polystyrene strands to be superior to apholate dusting (Patterson et al., 1972a).

The auto-sterilization principle has been applied to C. p. fatigans though not successfully as yet. Grant et al., (1970) fitted a UV-light to a CDC miniature light trap (Sudia and Chamberlain, 1962) and a tepa-treated plastic chamber into which mosquitoes were drawn by the suction fan and out of which they could escape into a collecting cage. Laboratory-reared males were fed into the trap and when they finally reached the collecting cage were removed and mated to virgin females. 90% to 100% sterility was achieved when the chamber was treated with 100 mg of tepa per square foot. The trap was then tried in the field in Florida but the light had to be removed and a baby chick added before *C. p. fatigans* were attracted and then only 8 out of 168 rafts derived from females captured over a period of 12 days were sterile. On this occasion treatment was at 60 mg of tepa per square foot.

The first field release of chemosterilized *C. p. fatigans* was made by Patterson *et al.*, (1970a) on the island of Seahorse Key off the coast of Florida in 1968. The island is a very small one (1·6 km × 0·2 km) only 3·2 km from the west coast of Florida and supported a population of *C. p. fatigans* in only a few man-made breeding places producing some 2 500 individuals of both sexes per day. Mass-rearing was done on the spot using wild-caught egg-rafts and sterilization was achieved by allowing emerging males (separated as pupae) to walk through plastic tubes previously dipped in 5% tepa in methanol and allowed to dry. Seven artificial oviposition sites were provided and recovered egg-rafts kept to see if they hatched. Daily releases varied from 2 250 to 2 750 and a total of 141 400 sterile males were released over a period of eight weeks (the months of August and September) representing some four generations in time. The calculated ratios of sterile to fertile males increased from 2:1 in the first generation of release to 5·7:1 in the fourth. Sterilities steadily increased to a final 85% but decreased to 51% in the generation after releases ceased and eradication was not achieved.

In 1969 (Patterson *et al.*, 1970b) male pupae were sterilized by immersion for 4 h in a 0·75% aqueous solution of thiotepa. This was done in the laboratory at Gainesville, Florida, and after treatment the pupae were transported on blotting paper in ice-chilled polystyrene boxes to a single release site on the island. Releases averaged 13 300 sterile males per day from June to August ranging from 8 400 to 18 000. By the tenth week after releases started there was a reduction of 99·8% in the number of fertile egg-rafts recovered and no wild larvae could be found. The analysis of data from these two field trials and the calculations of maximum net reproduction rates of a wild population of *C. p. fatigans* (Weidhaas *et al.*, 1971) has already been referred to in chapter two in the section on dynamics.

Another field release of chemosterilized *C. p. fatigans* was made in Kenya by Bransby-Williams (1971). Here two-day-old males were fed on 0·015% apholate in sugar solution to sterilize them and 32 350 were released over a period of ten weeks from three points in a village where the wild density was of the order of eight to twelve females per house in the first five weeks and fell to one per house in the tenth week. During most of this period, however, 77–100% of egg-rafts collected hatched.

Much larger-scale releases of chemosterilized C. p. fatigans have recently been made by the WHO/ICMR Delhi Research Unit on the Genetic Control of Mosquitoes. Preparatory laboratory observations (Sharma et al., 1973) have shown pupal immersion in 0·6% thiotepa for 3 h sufficient to give 99 ± 1% sterility in emerging males, though only 38% in females. The age of the pupae at the time of sterilization did not seem to affect the subsequent mating performance of the males which were equally, if not more, competitive than untreated males, both in laboratory cages and large outdoor cages. The mating ability of the males was assessed by caging them at a ratio of 1:6 with virgin females for five days and then replacing the females with further virgins for successive five-day periods. The sterilized males inseminated as many females as normal males in the first five-day period but were less efficient in successive periods despite little difference in longevity. That the accessory glands of treated males was unaffected was shown by replacing untreated males with treated ones and vice versa in cages containing females. 99% of those females caged first with untreated males continued to produce fertile egg-rafts while only 2–3% of those females caged first with sterilized males did so. No recovery of fertility of sterilized males could be detected up to 18 days after treatment. Field releases of differently-marked treated and untreated males and recaptures showed the treated males could fly as far from release sites and that they survived as well after release (Rajagopalan et al., 1973). Large-scale field releases were made in a village, Bamnauli, near Delhi, in 1971 (WHO, VBC/72·2).[1] The natural population was estimated to be in the region of 46 000 at the start of the trial. Pupae after treatment were placed in artificial containers at 25 sites within the village and about 80 000 sterile males were released on alternate days for four weeks. After this time 95% of 123 egg-rafts collected were sterile.

In 1972 further releases were made in the same village from July and in another village, Dhulsiras, from March.[2] Adults emerging from treated pupae were marked and released when 2–3 days old, giving time for some of the "contaminating" females to be inseminated by their own sterile males. In Dhulsiras releases started at about 10 000 sterilized males a day, rising to 40 000 by the end of May. During this time there was an eight-fold rise in the wild population size and the maximum sterility achieved was 22%. From June onwards, releases were increased to between 50 000

[1] World Health Organization mimeographed document (1972)—report of a technical planning and review group—VBC/72.2.
[2] Information from monthly reports of the WHO/ICMR Research Unit on the Genetic Control of Mosquitoes, Delhi.

and 100 000 per day and similar numbers were released in Bamnauli. They continued until the end of October in Dhulsiras and until mid-November in Bamnauli. From the end of September onwards the ratio of sterile to fertile males estimated from changes in the sex ratio in the villages varied from 12:1–88:1 and sterilities in recovered egg-rafts never exceeded 61%. These disappointing results gave rise to suspicions that the villages were insufficiently isolated and that wild, fertile females were entering them from outside. Thus barrier zones were instituted and preparations made for the stocking of breeding wells in them with mosquito-eating fish and the treatment of other breeding areas with insecticide.

In mid-February 1973, releases were restarted in Dhulsiras and combined initially with insecticide (fenthion) treatment of one of the village wells and a fogging of the whole village with malathion. An average of 150 000 sterilized mosquitoes, containing only 0·2% females and 1–2 days old were released each day until early May, followed by an increase to 300 000 per day for just over one month and a subsequent reversion to 150 000 per day. Results were disappointing to start with. With ratios of sterile to fertile males varying between 4:1 and 148:1 from catches made inside the village and surrounding areas and as high as 778:1 in swarms at dusk soon after releases were made, the proportion of sterile egg-rafts at the end of May was only 39%. During most of this period no evidence of adult emergence could be detected in the barrier zone and only limited breeding was taking place in Dhulsiras itself. What was happening, however, was that numerous normally-fertilized females were entering Dhulsiras from surrounding untreated areas. In addition, emigration was also occurring and sterile egg-rafts which could only have come from Dhulsiras females were encountered in villages more than 7 km away. In June mosquito immigration declined considerably and sterility in the egg-rafts increased to the much more satisfactory level of 87% towards the end of this month.

(iii) *Anopheline mosquitoes*

So far little has been done with chemosterilants against anopheline mosquitoes. Bertram and Giglioli (1963) made an isolated observation that a high sterility resulted in wild-caught. *An. melas* exposed for 1–3 h to 20–40 mg/sq. ft. of thiotepa in the Gambia. Schmidt et al., (1964) compared the competitiveness of *An. quadrimaculatus* males sterilized with apholate, tepa and gamma-rays. Males were allowed to feed on 1% apholate in sugar solution for the first three or four days of their lives or were exposed for two hours to deposits of 7 mg/sq. ft. of tepa. Sterilization by irradiation was achieved by exposing pupae more than 24 h old to 10 000 r and 12 000 r. Tepa proved better than apholate and better than 12 000 r. 10 000 r was as effective as tepa but not fully sterilizing.

In the chapter on sterilization by irradiation, reference to the unsuccessful attempt to control *An. quadrimaculatus* in the field by the release of irradiated males, was made. Failure was attributed to the use of a population which had been in colony for a very long period of time. Dame *et al.*, (1964) went on to compare the competitiveness of colony and wild material sterilized in the three ways referred to in the previous paragraph. Small-scale releases of sterile colony males and marked unsterilized, virgin females were made in the same area where the previous large-scale irradiation trial had been carried out. Females were caught after releases were made and their ovipositions observed for hatching. While only up to 13% of wild females laid sterile egg batches, up to 69% of the colony females did so, indicating that the colony males preferred to mate with their own females rather than the wild ones. The daily release of 900–1 400 tepa-sterilized colony males for two weeks produced no increase in sterility in the wild population and marked virgin colony females released half-way through this sterile male release programme showed 90% sterility. That this lack of mating between colony and wild mosquitoes was due to reduced mobility of the males was shown when F_1 females from wild material, reared in the laboratory, were released with tepa-sterilized colony males and virgin colony females. Then 50% of recaptured, laboratory-reared wild females were sterile as compared with only 4% of wild females; 93% of recaptured colony females were sterile. Finally, F_1 males from wild material reared in the laboratory and exposed to tepa deposits were released at the rate of 350 to 3 300 per day for 17 days. The sterility in the wild population increased from 0·23% to 1·10%, an increase of 4·8 times. Virgin laboratory-reared wild females and colony females released during this male-release programme showed sterilities of 15% and 23% respectively.

Large outdoor cage experiments with colony *An. quadrimaculatus* carried out by Patterson *et al.*, (1968b) showed that three-day-old males sterilized with apholate dust had a very short life-span and only a very small proportion of those released actually competed with normal males. Thus, theoretical ratios of up to 100 sterile to one fertile male were probably effectively nearer to 5:1 and the degree of sterility in females (80–90%) was that expected from this lower effective ratio. Here it was a question of a poor survival of laboratory-reared material in outdoor conditions.

A very recent field trial involving large-scale releases of chemosterilized *An. albimanus* is reported by Weidhaas *et al.*, (1973).[1] This took place between April and September 1972 at Lake Apastepeque in El Salvador. Pupae were sterilized in 1% aqueous solution of p,p-bis(1-aziridinyl)-N-

---

[1] In World Health Organization mimeographed document (1973) "Vector Genetics Information Service 1972" VBC/G/73.1.

methylphosphinothioic amide after a pupal-size separation allowing 14% females through. Male sterility averaged 99·8% and female 96·3%. Releases were made around the 3 km shoreline of the lake and in a nearby breeding area and were estimated at 30 000 a day, a total of 4·36 million over the five-month period. At the end of the trial the larval population reduction was more than 99·9% and one sterile female was all that was captured in the last two weeks of the trial.

(iv) Reduviid bugs

Among the other insects of medical importance *Rhodnius prolixus* has been the subject of experiments with the chemosterilant metepa. 0·1 mg per adult male topically applied to the thorax in acetone solution has been found to be the optimum concentration. It produces 98·8% sterility and only slightly affects mating competitiveness. It is superior to irradiation in this respect though some recovery in fertility occurs in time. Preliminary laboratory trials using plastic strips impregnated with a 25% solution of metepa have shown that nymphs and adults are slightly attracted to such strips and are sterilized by short contact with them. Containers of such strips, placed in infested houses resulted in nearly all the bugs in the house becoming sterilized in 29 days. As *R. prolixus* takes 83 days to complete its life-cycle, no single individual is likely to escape contact, according to Gomez-Núñez (1971).

Carcavallo and Carabajal (1971) have experimented with other reduviid bugs with tepa. Three species were used: *Triatoma infestans*, *T. guasayana* and *T. patagonica*. Tepa was applied topically to the fifth nymphal and adult stages and the nymphs proved more susceptible. Slight differences in the susceptibility of the three species were also apparent.

II. Agricultural Pests

A. Fruit Flies

Chemosterilants do not appear to have been used as frequently for the control of insects of agricultural importance. The Mexican fruit fly (*Anastrepha ludens*) was apparently equally successfully stopped from invading the fruit-growing areas of southern California by the release of tepa-sterilized flies as by irradiated ones but irradiation was eventually preferred to avoid toxicity hazards and because the simpler handling procedure of the irradiation technique resulted in increased fly vigour and survival (see Proverbs, 1969). In the chemosterilization procedure, pupae were dipped in a 5% aqueous solution of tepa for one minute and allowed to dry for 24 h (Shaw and Sanchez Riviello, 1965). Chemosterilants have been tried on other species of fruit flies, for example, tepa, apholate, metepa and tretamine on *Dacus dorsalis*, *D. cucurbitae D. oleae* and *Ceratitis capitata* (Keiser *et al.*, 1965; Orphanidis and Kalmoukos, 1971) but mostly in the laboratory. It appears to be a general rule that the male is more readily sterilized than the female.

B. The Cotton Boll Weevil

Davich (1969) describes field attempts to control the cotton boll weevil (*Anthonomus grandis*) by the release of males sterilized with apholate. Males had to be separated from females and then dipped in 2% apholate solution for 15 s on two occasions, 24 h apart. A small-scale release in a field in Louisiana, deliberately infested for the experiment, was an apparent success when a total of 8 850 males were introduced over a period of some six weeks. Initial calculations indicated a 20:1 ratio of sterile to fertile males, but the actual ratios were probably nearer to 400:1. No oviposition punctures could be detected at the end of the release period and no hatch of recovered eggs was recorded. Later larger releases combined with insecticide spraying were made in Alabama and involved the rearing of some 750 000 weevils to release 134 800 sterilized and surviving males at 19 700 per hectare over a period of eleven weeks. Some post-treatment mortality was evident even though the sterility was not complete and the surviving males showed reduced competitiveness. End of season data showed lower infestations in fields where releases were made than in those treated only with insecticide, but there was no eradication. It was concluded that the test fields were insufficiently isolated and that the population had not been reduced to a low enough level before releases started.

C. Lepidopteran Pests

(i) *The Red Bollworm*
It may be recalled that compared with most other insect pests, Lepidoptera require particularly high irradiation doses for complete sterility and that such doses have harmful effects on the insect's fitness. This is not the case with chemosterilants. Thus the injected doses required to produce 50% sterility in male house flies (average weight 15 mg) are 0·1 μg of tepa, 0·404 μg of apholate and 1·3 μg of metepa (Chang and Bořkovec, 1964). Equivalent doses for males of the African Lepidopteran *Diparopsis castanea* (red bollworm) (average weight 100 mg) are 0·73 μg of tepa, 1·76 μg of apholate and 2·81 μg of metepa (Campion and Lewis, 1971). Thus on a weight basis *Diparopsis* is as sensitive to tepa and apholate as the house fly but more sensitive to metepa. Campion and Lewis (1971) compared injection, topical and tarsal contact applications of hempa as

well as the three sterilants already mentioned to one-day-old *Diparopsis*. From injections they were able to establish a linear relationship between dose and resulting sterility and calculate the sterility index from the ratio of sterilizing dose to toxic dose. A high index is indicative of a wide margin between sterilizing and lethal dose, a necessary attribute of a good chemosterilant. Apholate proved to have the highest index (24·0) against male *Diparopsis* but affected mating ability. Tepa also had a high index (17·9) and did not affect male performance. Also the sterility was permanent. However, the index for females was quite low (4·2) and the dose required for high sterility affected both the mating ability and fecundity. Hempa had an extremely low sterility index (1·4). Tepa also had a high sterility index (34·6) when topically applied to males and the performance of such males, though affected on the first two days after treatment, was eventually nearly as good as untreated males. Residue analyses of tepa-treated moths showed rapid losses of injected compound but slower losses when topically applied due to slow adsorption. This is something of a drawback as topical application is the most likely technique for field release programmes though in this case the insect is a pest of cotton and not a food crop.

(ii) *The Cabbage Looper*

Tepa was also the sterilant studied by Henneberry (1971) in experiments with the cabbage looper (*Trichoplusia ni*). For complete sterility, injected doses of 85 μg for the male and 125 μg for the female were required. A comparison of feeds containing sterilant, spray applications, topical applications and tarsal contact with dry deposits, showed that those applications needed for high degrees of sterility affected the longevity and competitiveness of males and even their response to the female pheromone. These effects were less however when application was by spray. Howland *et al.*, (1966) had some success in controlling this species in cages by combining chemosterilant and light attractant. Fifteen-watt fluorescent black-light lamps were enclosed in cellulose nitrate cylinders, the outer surfaces of which were coated with an 8% tepa solution. Moths released into cages containing such lamps were apparently sterilized when left for 14 consecutive nights and larval populations were 65 to 99% less than in control cages. Efforts to speed-up the intake of sterilizing doses from surface deposits were made by incorporating tepa in mixtures of lanolin and n-hexanol. 2-second contacts with 16% tepa in 75% lanolin were sufficient to sterilize both sexes. This was then used in conjunction with the black-light lamps and a high degree of sterility was produced in just one night's operation of the self-sterilizing apparatus (Henneberry *et al.*, 1965).

(iii) *The Pink Bollworm*

The pink bollworm (*Pectinophora gossypiella*), another cotton pest, has been controlled in large-cage experiments with metepa-sterilized males. A topical application of 15 μg of the sterilant in acetone to males less than 24 h old produced equally competitive but sterile insects which when caged with untreated males and females at a ratio of nine sterile to one normal male and one normal female resulted in the production of 81% less larvae than the control cages (Ouye *et al.*, 1965).

(iv) *The Codling Moth*

Attempts to control the codling moth (*Laspeyresia pomonella*) by the release of irradiated insects in both Canada and the U.S.A. were referred to in the previous chapter. The American field trials started in 1964 with tepa-sterilized males (Butt *et al.*, 1970). At first, newly-emerged males (less than 18 hours old) were dipped in a 2% aqueous solution of tepa but this inactivated them. In 1965 a topical application of 15 μg per male was tried but the results were inconclusive as far as reductions in fruit damage were concerned. For most of the release period the proportion of sterile to normal males was only 5:1 though it did reach 127:1 at the end of the trial. In 1966 sterilization was by exposure to an aqueous aerosol of 20% tepa for 10–15 min. but half-way through the season a change was made to gamma-irradiation at 40 000 r. An average of 271:1 of sterile to normal males was achieved in this trial and reductions in fruit damage were encouraging but whether this success can be attributed to chemosterilant or irradiation is uncertain.

D. Chemosterilant-Bait Combinations

Three instances of limited trials with chemosterilant-cum-bait combinations against agricultural pests can be cited. Orphanidis *et al.*, (1966) used bait stations containing 0·4% apholate against the olive fruit fly (*D. oleae*) in an isolated olive grove on the island of Lesbos. Egg hatches from females caught in this grove for a period of two and a half months were consistently less (about half) than in a control grove. Luckmann *et al.*, (1967) similarly used apholate bait stations in large-cage tests for the control of the onion maggot (*Hylemya antiqua*) producing more than 89% sterility. Finally Coaker and Smith (1970) produced 70% sterility in the cabbage root fly (*Erioischia brassicae*) when food baits containing 0·1% tepa, sucrose and leucine were introduced into large field cages.

5. Hybrid Sterility

A species of animal or plant is a population of individuals which will interbreed but even when temporarily or spatially contiguous with populations of other species, remains reproductively isolated from them. The isolating mechanism in animals takes the form of a mating barrier which may involve differences in sexual signals of sound or pheromone or differences in time, place or type of mating activity. This barrier rarely breaks down in nature so that natural hybrids are uncommon. However, its breakdown can be brought about in certain artificial conditions.

Matings between species can take the form of attempted copulations, where the genitalia do not connect properly because of structural differences, actual copulations without insemination and copulations with insemination. Insemination may or may not be followed by actual fertilization of the ova by the spermatozoa. If such fertilization does occur the hybrid individual may then die in the embryonic stage (before escape from the egg-shell or uterus), in the pre-adult stages (early or late) or prematurely in the adult stage, or the hybrid adult may be a perfectly normal individual from external appearance, behaviour and survival, even showing evidence of hybrid vigour or heterosis. It will, however, normally show some abnormality in gametogenesis and partial or complete sterility in one or both sexes. Which one of these states results—no progeny, premature F_1 death or normal or heterotic sterile adult is usually taken, perhaps wrongly, as an indication of the phylogenetic proximity of the species involved, the more closely related species having the greater likelihood of producing hybrid adult offspring.

Hybrid vigour can take the form of increased size, increased survival from egg to adult, increased adult longevity and even increased sexual activity and competitiveness under some conditions. The strength, stubbornness and stamina of the mule (the hybrid produced by crossing a male ass with a female horse) are well known. What is perhaps less well known is the sexual aggressiveness of this animal which leads breeders to isolate it from the parent stocks of horses and asses. It is therefore conceivable that a hybrid between two species may not be reproductively isolated from one or both of its parents. If this is the case a potential method of genetic control exists dependent on the mass production and release of sterile hybrids. Such a method will not involve costly or dangerous sterilization methods or complicated breeding techniques, only the ability to mass-rear two species and to cross them.

I. Anopheline Mosquitoes

Anopheline mosquitoes provide us with good examples of hybrid sterility. Crossing can often be readily achieved by a forced-mating technique between single pairs or through mass cage-matings using colonies adapted to laboratory-cage conditions. The artificial mating technique involves the presentation of the genitalia of an anaesthetized female to those of a decapitated male. Using this technique it has been possible to cross species which will not mate in cages.

A. The Anopheles maculipennis Complex

The first cases of hybrid sterility in anopheline mosquitoes were demonstrated in what is now known as the *Anopheles maculipennis* complex and are described by Hackett (1937). Seven varieties were recognized in Europe each having a distinctive type of egg, though the adults were indistinguishable from one another. One of these, variety *atroparvus*, would readily mate in small cages and the males would readily mate with the females of the other varieties. Five crosses were carried out. Two of them produced larvae which died, in one case early and in the other at an advanced stage. One produced adults with both sexes sterile. Another produced adults in which all the males were sterile and about half of the females. The fifth produced all fertile females and some fertile males. These crosses were performed before an artificial mating technique was developed. More than twenty years was to elapse before the reciprocal crosses could be made by this method and again variations in the survival and sterility of the F_1 generations were apparent (Kitzmiller *et al.*, 1967). Variations in the sex-ratio of the F_1 generation were also shown usually with distortion in favour of the female sex.

The importance of this work to Hackett and his contemporaries was the clear demonstration that they were dealing with a number of different types which though indistinguishable except in the egg-stage, and then only with difficulty between some of them, were definite species each with their own characteristics. The characters which concerned them most was their ability to transmit malaria and it was shown eventually that some of them did not transmit the disease and could thus be ignored in control activities.

Other workers, notably G. Frizzi and J. B. Kitzmiller (see Kitzmiller *et al.*, 1967) later went on to study the chromosomes of these species and other relatives in the New World and showed differences in the banding sequences of polytenes sufficient to enable species identification in most cases.

Hybrid chromosome preparations showed numerous non-homologous areas (areas of asynapsis) between "homologous" chromosome pairs and these incompatibilities could account for hybrid sterility. Now the palaearctic *Anopheles maculipennis* is recognized as a complex of five sibling species (and two subspecies). No attempt has ever been made to use sterile hybrids from this complex for control purposes, primarily because malaria has now been largely eradicated, by the use of insecticides, from areas where the species occur, in Europe at any rate.

B. The Anopheles gambiae Complex

(i) *General*

The second species complex to be studied was that of *An. gambiae*, the notorious African malaria vector. Until 1956 this was considered a single species with a dark, coastal, salt-water variety *An. melas* recognized as occurring in West Africa and a morphologically indistinguishable East African salt-water form *An. merus* occurring along the East coast and on islands off it. In 1956 crossings were being made in London between two populations of fresh-water *An. gambiae* from Nigeria, one a dieldrin-resistant population from Western Sokoto, Northern Nigeria and one an insecticide-susceptible population from Lagos in the course of a study of the mode of inheritance of dieldrin-resistance (Davidson, 1956). A hybrid generation was produced from reciprocal mass-crosses of cage colonies but attempts to derive an F_2 generation from them failed. Masses of eggs were laid but none hatched. The females were proved to be fertile by backcrossing them to the parent males and obtaining progeny. It was the F_1 males that were at fault and an examination of their testes showed them to be atrophied.

Since then numerous crosses of material from all over the African continent and Arabian Peninsula and from many of the islands, e.g., Mauritius, Madagascar, Zanzibar and Fernando Poo, including material from salt-water breeding sites, have shown *An. gambiae* to be a complex of at least six species all producing F_1 male abnormality (not full sterility in some cases) when crossed (Davidson, 1964; Davidson *et al.*, 1967; Davidson and Hunt, 1973).

Three of the six species are fresh-water forms provisionally called species A, B and C. The first two are highly efficient malaria vectors occurring throughout the African continent south of the Sahara and often together. They have been found in the same breeding place and resting in the same house. Species C so far, has only been found in south-east Africa (South Africa, Mozambique, Swaziland, Rhodesia and Zambia), Ethiopia and the islands of Zanzibar and Pemba. It appears to be entirely zoophilic and not a vector of malaria. Species D has only recently been identified from a forest population in Bwamba County, Uganda, where it

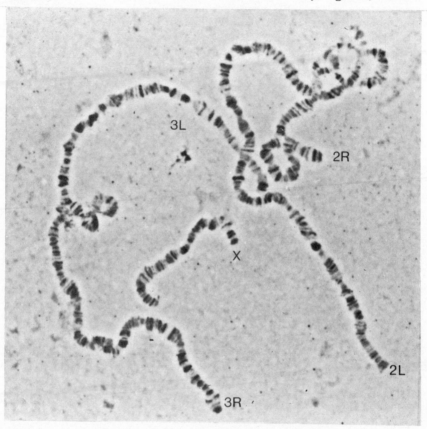

2R, 2L, 3R, 3L—autosomal arms   X=X—chromosome.

Fig. 4   Polytene chromosomes from the salivary gland of the fourth instar larva of species B of the *An. gambiae* complex.

breeds in spring water with a high content of various inorganic salts. Its vectorial state is unknown but it can and does exist in the absence of man. The other two species are salt-water species usually only found near the coast. *An. melas* is restricted to West Africa where it may be an efficient malaria carrier, near the centre of its distribution at least. *An.*

merus occurs in East Africa where it certainly carries malaria but its exact epidemiological importance is uncertain.

Certain identification of the member species of this complex from conventional morphological features has proved impossible to date. More than 350 characters in all stages of the life-cycle have been examined yet no-one has produced a means of identification of single individuals from such characters. The success of the use of polytene chromosomes in the identification of the member species of the *An. maculipennis* complex prompted a detailed examination of these in the *An. gambiae* complex and proved equally successful. Now all six species can be positively identified

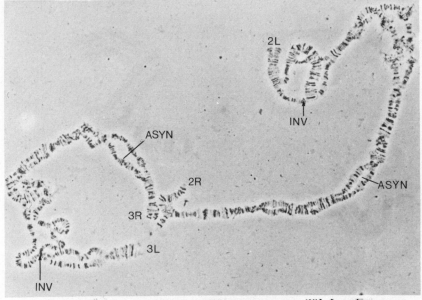

2R, 2L, 3R, 3L—autosomal arms   ASYN—asynaptic area   INV—inversion

Fig. 5   Polytene chromosomes from the nurse cells of the ovary of a hybrid between species A and species D of the *An. gambiae* complex.

from differences in the banding sequence of the chromosomes particularly that of the X-chromosome. Early work was confined to the polytene chromosomes of the salivary glands of the fourth instar larvae (Fig. 4). Later such chromosomes were discovered in the nurse cells of the half-developed ovary and now direct identification of the adult female mosquito is possible (Coluzzi and Sabatini, 1967, 1968a, 1969; Coluzzi, 1968; Green, 1972; Davidson and Hunt, 1973). The examination of chromosome preparations from hybrid mosquitoes has greatly assisted in the recognition of these chromosome differences, these differences causing asynaptic areas (Fig. 5).

Insecticide resistance is already widespread in species A and B though it has not yet been found in the other four species. Dieldrin resistance is particularly common in West Africa even in areas where insecticides have never been used. This resistance has also been found in the same two species in Kenya, Sudan and Madagascar in the East. The use of dieldrin has now been virtually abandoned throughout the continent and its place taken by DDT. Now, however, resistance to this insecticide is appearing in Upper Volta, Togo, Senegal, Sudan and South Africa and again is present in both species A and B. Apart from insecticide resistance and indeed from technical problems in general there are many other difficulties in the way of control of malaria in Africa, the most important being largely economic ones. Without going into details suffice it to say that alternative control methods to the ones used at the present time will be essential if any progress is to be made in the control of malaria on this continent.

While sympatric associations of two or more species of this complex are common, hybridization is rare in nature. Paterson (1964), for example, found no evidence of sterile males in the offspring of 174 females caught at Chirundu, Rhodesia, where species A, B and C had been identified. Ramsdale and Leport (1967) found only one family showing all sterile males from 202 ovipositions from females caught in a suburb of Ibadan, Nigeria, where species A and B occurred, the former predominating. White *et al.*, (1972) examined a total of 5 196 chromosome preparations in an area of Tanzania where species A and B occurred together and changed in predominance from one part of the year to the next. Altogether eight hybrids were recognized and all but one were found when one or other species predominated. Thus an occasional breakdown in species mating barrier does occur and is most likely when one species swamps the other.

Within six species there are thirty possible crosses. Twenty-nine of these have been made, either by artificial mating or by mass cage-crossings. Mass crosses in small cages (as small as $8 \times 8 \times 8$ in.) have usually succeeded when the parent colonies have adapted themselves to laboratory conditions. What normally happens in laboratory cages is that wild populations straight from the field produce very few eggs at first, i.e., only a few individuals contribute to the continuation of the population. These almost certainly are those individuals genetically endowed with the ability to mate without swarming, the normal prerequisite to mating in most mosquitoes. Their offspring inherit this ability and after a difficult start the newly acquired laboratory colony becomes easier and easier to maintain until after a time (which may vary from several weeks to several years) a situation may be reached where quite high fertilization rates are achieved within a week, or even less, of emergence. This process, then, would appear to involve a gradual selection for those individuals which mate in small spaces and the final production of a colony or population whose mating behaviour is different from the population from which it was derived. In fact it is possible to reach the stage where on adding one established colony of one species to another established colony of another species random mating may occur (Davidson, 1964), a very different situation from the sympatric existence without significant hybridization that normally occurs in the wild. The question that can really only be answered by a field release is whether hybrids from such established colonies (and these would have to be used for mass production) can recognize their wild, remote relatives and mate with them.

No sign of any hybrid mortality or weakness is evident from any of these crosses. Egg-hatches and F_1 yields are normally as high as from homologous matings except in 6 of the crosses, those between species A and B males and *An. melas*, *An. merus* and species C females. Here the male sex often predominates in the F_1 generation and this situation is brought about by the non-hatching of some of the eggs. Where there are no females at all the egg-hatch is always 50% or less and it thus appears that the X-bearing spermatozoon fails to unite with the egg nucleus. As already pointed out in Chapter 2 this "built-in" sex-separation is not a stable condition unfortunately, and it is important to limit the release of hybrid females which are not only capable of biting and transmitting disease but also, because of their fertility, of mating with wild males and producing progeny. In addition by mating with the hybrid males they will absorb some of their sterilizing effect.

The degree of atrophy of the reproductive system of the hybrid male and its degree of sterility depend on the parents involved. In some sterility is absolute and spermatogenesis has not gone beyond the round-cell stage of spermatocyte or spermatid, e.g., species B male × species A female, while in others tail-formation has proceeded in varying degree perhaps even to the extent of producing a few motile spermatozoa, e.g., species A male × species B female and species B male × *An. merus* female. In others, for example, species A male × species C female and species D male × species C female the hybrid male is near-normal in its fertility even though the reciprocal crosses are perfectly sterile. Finally, in some cases both testis and accessory gland are atrophied both to the extent of not producing spermatozoa and to the extent of not inducing the monogamous response in the female (see Chapter 2). Figures 6 and 7 show normal and sterile male reproductive systems.

(ii) *Laboratory Competition Experiments*
Numerous laboratory experiments have been made to test the competitiveness of these sterile hybrid males (Davidson, 1964, 1969a, 1969b;

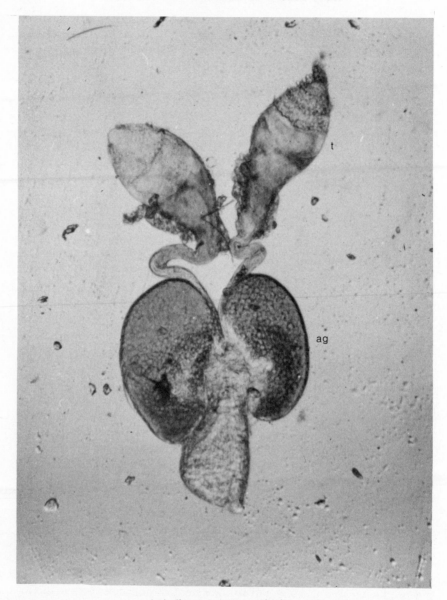

t=testis ag=accessory gland.

Fig. 6   The normal male reproductive system of *An. gambiae* species D.

Davidson *et al.*, 1967). The early ones involved the addition of adult sterile males in varying proportions to small cages containing normal males and normal females. The mixed adults were left together for some

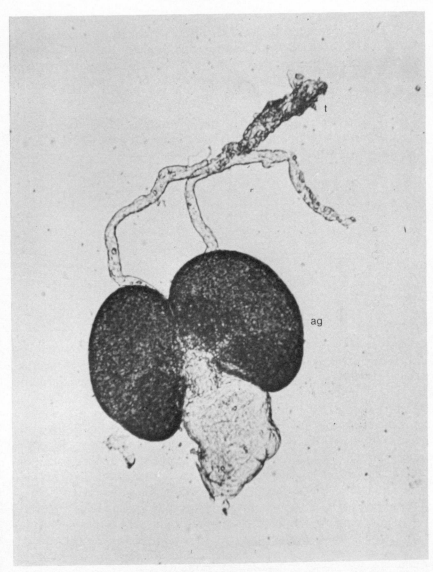

t=testis   ag=accessory gland.

Fig. 7 The sterile male reproductive system of a hybrid produced by crossing species D and species A of the *An. gambiae* complex.

days, the females fed on blood, tubed singly over water and subsequent ovipositions kept to see if they hatched. The sterile males used were produced from crosses between species A and species B and between species A male and *An. melas* female, species B male and *An. melas* female and species B male and *An. merus* female. They were tested against both species A and species B populations. Table II summarizes all the results from these adult additions. A comparison of expected and actual hatching of ovipositions shows a superior mating ability of the sterile males at all the ratios used from $\frac{1}{4}$:1 to 4:1.

TABLE II   The effect on the fertility of the females of adding sterile hybrid males to cages containing normal males and normal females.

Actual male proportions Sterile : Normal			Ovipositions Total	% hatching	Expected %
0	:	1	308	89	
$\frac{1}{4}$	:	1	132	70	71
$\frac{1}{2}$	:	1	133	52	59
1	:	1	339	34	45
2	:	1	362	23	30
3	:	1	179	17	22
4	:	1	133	17	18

Later laboratory experiments involved the F_1 generations of those crosses between fresh-water and salt-water species known to produce mainly male offspring. It was reasoned that a considerable saving in time and effort could be achieved if such F_1 generations were released as eggs into breeding places. Then the larvae could grow side by side with the normal ones and emerge at the same time and in the same place as the normal adults. In this way they would probably be better adjusted to local conditions than those reared under artificial laboratory conditions and in a better position to compete with the normal males. In the laboratory this situation was simulated by adding newly-hatched first-stage larvae from such crosses to larval rearing bowls containing first-stage larvae of species A or B, in varying proportions. The mixed larvae were then reared to the adult stage and the adults left together in cages for some days, as in the previous experiments. Again the numbers of sterile and fertile ovipositions were determined. Table III summarizes all the results from a series of such larval seeding experiments.

The crosses involved were between species A and B males and *An. melas* females and between species B males and *An. merus* females. The results again show significant changes in the percentage of fertile ovipositions caused by the presence of sterile males, though these changes are not as great as in the adult series of experiments. Male dissections made at the end of the experiments revealed that the actual final proprtions of sterile

TABLE III The effect on the fertility of the females of adding first-stage hybrid larvae (known to produce males which are sterile) to bowls containing normal first-stage larvae and rearing through to the adult stage.

Expected male proportions Sterile : Fertile			Total	Ovipositions % hatching	Expected %
0	:	1	592	85	
1	:	1	432	48	43
2	:	1	928	39	28
4	:	1	206	26	17

to normal males were in some cases less than the intended and this was accounted for by the sporadic occurrence of significant proportions of hybrid females in the F_1 generation of the crosses employed. In those experiments where fewer hybrid females could be detected, the actual ratio of sterile to normal males often exceeded the intended. This indicated an enhanced survival of the hybrid and deliberate rearing under adverse conditions followed by immediate and delayed male dissections showed that this enhanced survival was shown by both hybrid larva and hybrid adult.

Fig. 8 The village of Pala, Upper Volta.

(iii) *Field trial*

This was as far as one could go in the laboratory. The next step was to see if these sterile males were as competitive in the wild as they so obviously

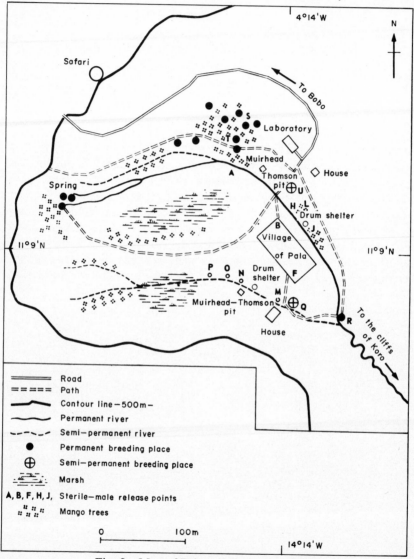

Fig. 9   Map of Pala and its surroundings.

were in laboratory cages. For the field trial a small village called Pala in Upper Volta was chosen (Fig. 8). This had a population of some 500 people living in some 300 rooms and from the beginning of the dry season onwards could be considered as free from mosquito immigration. The sole species of the *An. gambiae* complex was species A which from October onwards, after the rainy season, declined markedly in density to a very low level in the month of December and remained at this level until the start of the following rainy season, usually in April. Sterile male releases were planned to coincide with the population decline and to be completed before the marked temperature drop in mid-December characteristic of that part of the world. The trial took place in 1968.

The original intention was to seed the few breeding places left in Pala

Fig. 10 Releasing sterile male pupae into natural, temporary, breeding place M at Pala.

at the beginning of the dry season with eggs (when they were about to hatch) from a cross known to produce mainly sterile-male offspring. Ideally the cross would be between the male of the natural population, viz., species A and female *An. melas* (this being the West African salt-water species). Trials of this cross in London had, however, always resulted in the production of high proportions of females in the F_1 generation. The most successful cross in the London laboratory experiments had been one between a species B population derived from Kano, Nigeria, in 1958 and an *An. melas* population derived from Harbel, Liberia, in 1962. This cross resulted in fewer females in the F_1 generation and the males were highly competitive with normal males. It was decided therefore to transfer these two colonies to Upper Volta for the large-scale production of sterile males. Though it was realized that a cross between two species was being used to attempt to control a third species, it was thought possible that hybrid vigour might override species mating barriers.

The scientific application of this egg-seeding method depended on having some knowledge of the size of the natural aquatic population. Cuellar (1969a) had calculated that to eradicate a daily emerging population of 10 000 males and females would require the introduction into the

Fig. 11   One of the permanent breeding places (R) near Pala village.

breeding places of some 2 000 000 hybrid eggs a day for nine weeks. These calculations were based on the assumption that the sterile males were as good at mating as the natural ones and that no natural female would be expected to live longer than nine weeks. This number of hybrid eggs represented a ratio of two "factory" eggs to one natural and allowed for the expected 50% non-hatching of the F_1 eggs (this accounting for the sex-distortion, only male-producing eggs normally hatching). The calculations also took into account the fact that the mortality occurring between the egg stage and the emergence of the adult under field conditions must be exceedingly high (more than 98% in the static population referred to) and that the continuous addition of large quantities of hybrid material would ensure a continuing high aquatic-stage mortality.

The ideal situation of a few circumscribed breeding places which could be easily sampled by the traditional mark–release–recapture ecological technique of population-size estimation did not exist, however. Late rains had resulted in much more breeding than expected and this for the most part in places impossible to sample (the edges and backwaters of a river, for example). For this reason and because it seemed doubtful whether the insectary would be capable of producing enough hybrid eggs, it was decided to rear the F_1 generation to the pupal stage under laboratory conditions with only a moderate mortality and to release the pupae in and around the village. Cuellar (1969b) had calculated that only 300 000 sterile males per day would be sufficient to eradicate a population maintained by 10 000 daily emerging males and females if released over a period of thirteen weeks. The technique used in the production of this F_1 generation have been described in Chapter 2.

The pupae were transported on damp gauze in Petri-dishes to the release sites. The accompanying map (Fig. 9) shows the sites in and around the village. For about the first month, most of the releases were made into natural breeding places showing heavy densities with high proportions of advanced aquatic stages, e.g., points L and M (Fig. 10). Masses of pupal skins in these breeding places indicated good emergences. However, towards the end of November the main breeding places dried up and a sudden fall in night-time temperature to as low as 12°C occurred. Releases were then made from artificial containers, both immersed in natural water collections around the village and left in empty houses in the village, and also from a few permanent pools showing light breeding at some distance from the village houses, e.g., points R, S and T (Fig. 11). High mortalities were evident from such releases and eventually all releases were made from glass jars bedded in straw in predator-proof cages at point J (Figs 12 and 13).

After releases started, routine hand-catches were made of both males and females resting in houses and outside pit-shelters in the daytime. The males were dissected and their testes examined to see if they were fertile or sterile while the females were tubed individually over water for oviposition. Those females not developing eggs were dissected and classified as fertilized or not (from an examination of the spermatheca) and nulliparous or parous (from an examination of the tracheal system of the ovary). Gravid females dying without ovipositing were also dissected to see if their spermathecae contained spermatozoa. Ovipositions were kept for several days to see if they hatched and were classified as normal in size and floating, small and sunken. In addition samples of eggs from the ovipositions were examined to see if they came from natural species A females or to see if they came from hybrid females. Eggs of the latter resembled those of *An. melas* having a broad deck and no significant gap

Fig. 12   Predator-proof release cages at point J, Pala.

Fig. 13   Jars containing sterile male pupae bedded in straw in a release cage.

between the dorsal edge of the float and the frill of the deck. *An. gambiae* species A eggs usually have a narrow, waisted deck and a considerable gap laterally between float and frill. Females which oviposited and survived and ovipositions which hatched were all returned to the village.

In the control villages only house-catches were carried out and no return of captured mosquitoes was made.

TABLE IV   Gross results of sterile male releases.

	Pala	Controls
Pupae released	295 813	
% mortality	10 (56 651)	
% female	7 (21 676)	
% of males captured which were sterile:		
from houses	66 (275)	
from outside shelters	83 (173)	
No. of females ovipositing	357	296
No. of ovipositions hatching	319	276
Non-hatchers:		
Normal—species "A"	12	4
Normal—hybrid	9	0
Abnormal	17	16

(Figures in brackets are the actual numbers from which the percentages were calculated)

Full details of the trial are given by Davidson *et al.*, 1970, and Davidson, 1971. Table IV summarizes the results in the treated and control villages while Table V attempts to relate in approximate weekly periods actual releases of sterile males to those calculated to be required from density estimates. Density estimates have been made assuming that the hand-catching method is 50% successful and those mosquitoes resting in houses represent one quarter of the total (the remaining 75% resting outdoors). The natural daily survival rate of the female has been taken as 0·9 and from this it is calculated that the average number of ovipositions per female is three in a period of ten days. The number of rooms in Pala is taken as 300. Thus five females per room

$$= \frac{5 \times 2 \times 4 \times 300}{10} = 1\ 200 \text{ females}$$

$= 2\ 400$ males and females per day.

The required number of sterile males is calculated from Cuellar's (1969b)

TABLE V Actual and predicted results of sterile male releases by intervals.

	Females per room per day	Calculated daily emergence (male and female)	Sterile males required per day	Actual sterile males emerged per day	% of males captured which were sterile	Pala normal species "A" layings which did not hatch	Control normal layings which did not hatch
10/11–16/11	5	2 400	72 000	2 600	—	—	0/40
17/11–23/11	2	960	28 800	4 000	68	3/61	1/41
24/11–30/11	1·3	624	18 720	5 900	76	3/89	0/33
1/12–7/12	1	480	14 400	6 000	85	2/35	2/39
8/12–14/12	0·4	192	5 760	5 700	86	3/60	0/27
15/12–21/12	0·3	144	4 320	4 500	54	0/10	0/41
22/12–28/12	0·1	48	1 440	2 500	87	0/4	—
29/12–3/1	0	?	?	2 300	89	—	0/21
4/1–13/1	0·02	10	300	?	61	0/2	—

300 000 per day necessary to eradicate 10 000 daily emerging adults of both sexes. The actual number of sterile males is taken as 80% of the pupae released and allows for a mortality (which varied from 5%–16%) and for a proportion of females (which varied from less than 1%–11%).

It can be seen from Table V that only from the second week in December did the number of sterile males released reach the estimated number required and this stiuation only continued for about one month. However, for more than six weeks 75% of the males captured from houses and from outside shelters were the introduced sterile males. From the beginning of December the density of the natural population declined rapidly, until at the end of the month only an occasional female could be recovered by the hand-catching technique. In the main control village, however, adult densities were higher and though declining were still at two females per room at the end of the trial.

A total of 1 804 females were tubed from the village of Pala before the sterile male releases took place, and from two control villages over the whole period. 70% of them oviposited. In Pala, after the releases started, 70% of 502 females laid eggs, 23% died in the gravid state and 7% did not develop eggs. Most of the latter were unfertilized and three-quarters of them were nulliparous. Similar proportions were obtained with females caught in the control villages. Of 1 254 ovipositions obtained from Pala before releases were made and from the two control villages over the whole period of the trial, 1 214 hatched. Of the 31 ovipositions which failed to hatch, 22 were considered as abnormal. After releases started, the percentage of abnormal layings was about the same (around 5%) in both the treated village of Pala and in the control villages. The total percentage of what were considered to be normal layings which did not hatch was 5·88 from females caught in Pala after some two months of sterile male releases. In the same period in the control villages 1·35% of normal-looking layings did not hatch. However, 2·52% of the normal-looking ovipositions from Pala proved to be from hybrid females and only 3·36% from species A females.

Various reasons can be sugggested to explain the disappointing results of this field trial. Perhaps the three principal ones are:

(1) Adverse climatic conditions at the time of peak release affected the mating competitiveness of the introduced sterile males. Low night-time temperatures suddenly occurred leading to considerable mortality among introduced pupae not protected from the abrupt temperature change and delaying the dispersal of the emerged sterile males from the release points.

(2) Insufficient numbers of sterile males were released over too short a period of time. This is undoubtedly true. Prior to the fourth week in November quite high emergences of the natural species A population were in evidence. After this time the main breeding place dried up. It was not

until the second week in December that releases were anywhere near the numbers calculated as necessary. Thus a high emergence with a high proportion of matings between normal males and normal females could have produced a future masking of sterile male matings by the continued survival of fertile females (fertilized before adequate sterile male releases were made) against a background of a marked decline in new additions to the population. Such a situation has been commented on by Cuellar (1970) and has been referred to in Chapter 2 in the section on dynamics.

Cuellar (1973b) has since modified his computerized mathematical model to allow for the rapid drop in density (from 6·8 females per room to 0·4 in 5 weeks) instead of considering the population as a static one and has changed some of his parameters to more realistic values in the light of actual happenings at Pala. The maximum length of life of a female has been extended to 32 days instead of 30 and could perhaps be extended even further. The maximum probability of aquatic survival from egg to adult has been drastically reduced from 0·9 to 0·005, considered more representative of happenings in the few low-density permanent breeding places persisting in the dry season at Pala. This figure of 0·005 would give a potential rate of increase of less than five times. Changing all these parameters and allowing for the fact that the population was a rapidly declining one and introducing a partial mating barrier (eight sterile males as competitive as one normal one) Cuellar (1973b) found he could simulate very closely the actual proportions of sterile ovipositions obtained in Pala if the population was challenged with 4 000 sterile males a day. Further, if this challenge was continued for a further fourteen weeks from the time the females-per-room index reached 0·4 eradication would be achieved.

(3) A species mating barrier existed between sterile males and wild females and was exaggerated by the use of a cross between two species against a third. The fact that nearly half (9/21) the sterile layings in Pala were the result of matings between hybrid male and hybrid female, that no successful mating was ever shown to occur between natural male and hybrid female (in other words no batch of hybrid eggs ever hatched) and the few sterile species A ovipositions, all strongly suggest the existence of a mating barrier.

(iv) Post-trial Observations

Since the field trial a number of laboratory experiments have been made to try to assess the importance of mating barriers. One way in which a mating barrier between species might express itself is in differing times of mating activity. One element of mating activity is flight activity itself and an apparatus for the automatic recording of the sound of this activity has been developed by Jones (1964). Using this apparatus Cubbin (personal communication)[1] has recorded the patterns of activity of species A, species B, *An. melas*, *An. merus* and hybrids between them under conditions of constant darkness so that their "free-running rhythm" could be observed. In most cases four successive daily cycles of activity were recorded and from the records determinations of period (time between successive peaks of activity), latency (time from normal light-off time to first activity) and quantity of activity (minutes of activity in a 24 h period) have been made. The experiments showed no major differences between species A, species B, *An. melas* and the hybrids from crosses between species A male and *An. merus* female, species B male and *An. merus* female, species A male and *An. melas* female and species B male and *An. melas* female. *An. merus* showed some differences at least in the period of activity and there was some evidence of a higher degree of activity in this species and in some of the hybrids with it (perhaps an expression of hybrid vigour). However, *An. merus* was not used in the Upper Volta trial.

Another reason for the maintenance of identity of sibling species even when they occur together may be a differing response to sounds. Using the same acoustic apparatus used to study endogenous rhythms of flight-activity Cuellar (personal communication)[2] recorded the flight sounds of 4 of the 6 species of the *An. gambiae* complex. These have been played to cages of mosquitoes at the appropriate evening activity period and definite responses of males to the sounds have been demonstrated. These responses are far from species specific however, and preliminary quantitative sound-attraction experiments indicate a greater response of the hybrid male produced from the cross between species A male and *An. melas* female to the recorded sound of a species A female than that of the species A males themselves.

The mating competitiveness of a number of different hybrid sterile males has been tested since the field trial under standard conditions in the laboratory and against the same populations of species A and species B as were used in the crosses to produce the hybrids, viz., the species A population from Pala, Upper Volta and the species B from Kano, Nigeria. The *An. melas* population, however, was a different one from the one used in the field trial and came from the Gambia instead of Liberia. The *An. merus* population originated from Tanzania. A ratio of two sterile to one normal male was used, the mixtures of adults left in $8 \times 8 \times 8$ in. cages for one week and the fertilization of females assessed by microscopic examination of the spermatheca. Initial experiments were with cages containing 40 hybrid males, 20 normal males and 20 normal females.

[1] Miss C. Cubbin, formerly of Brunel University, London, and now at the College of Agricultural Sciences, University of California, Berkeley, 1050 San Pablo Avenue, Albany, California 94706.
[2] Dr. C. B. Cuellar formerly of the London School of Hygiene and Tropical Medicine. Present address: Apartado 77, Matamoros, Tamaulipas, Mexico.

Control cages contained only 20 normal males and 20 normal females. The results of these experiments are given in Table VI. The order of sterilizing efficiency of the various hybrid males based on the difference between actual fertilizations rates obtained and those expected assuming random mating were as follows:

Order Number	Sterile male derivation Male parent	Female parent	Target species
1	A	*merus*	A
2	A	*merus*	B
3	B	A	A
4	A	*melas*	A
5	B	*melas*	A
6	A	*melas*	B
7	B	*melas*	B
8	B	*merus*	B
9	A	B	A
10	*merus*	A	A
11	B	*merus*	A
12	B	A	B
13	*melas*	A	A
14	A	B	B

These results show that the hybrids used varied considerably in their efficiency and that in most cases involving crosses between species A and B males and *An. melas* and *An. merus* females, the hybrid is more efficient against one of its parents than against a third species. The cross used in the field trial (species B male × *An. melas* female) rated only a mediocre performance. The cross between Pala species A male and *An. melas* female proved more efficient and might be the favoured choice if a repeat field trial at Pala were to take place. The fact that the hybrid from the cross between species A male and *An. merus* female came top of the table may be a reflection of the higher activity of *An. merus* and hybrids with it shown in the endogenous rhythm observations described earlier. The hybrid male derived from the cross between species B male and species A female proved to be highly efficient against species A. Unfortunately this cross produces a normal sex ratio but if some efficient method of sex separation was available it might be given consideration for future trials. The hybrid males produced by crossing *An. melas* and *An. merus* males with species A female showed low efficiencies. These crosses pro-

TABLE VI Comparative efficiency of various hybrid sterile males.

Sterile male derivation male	Sterile male derivation female	Target species	No. of experiments	Percentage male mortality experiments	Percentage male mortality control	Percentage female mortality experiments	Percentage female mortality control	Fertilizations in experimental cages observed	Fertilizations in experimental cages expected	Ratio observed/expected
A	B	A	9	15	23	37	25	60/111	26·1/111	2·30
A	B	B	4	22	35	54	53	25/111	3·3/111	7·58
B	A	A	6	18	21	20	14	19/96	18·5/96	1·08
B	A	B	7	24	51	49	53	25/63	8·2/63	3·05
A	melas	A	15	17	33	25	14	78/238	59·8/238	1·30
A	melas	B	11	29	42	55	49	36/100	18·8/100	1·915
B	melas	A	12	17	10	28	12	87/176	45·6/176	1·908
B	melas	B	8	18	36	48	53	37/92	18·5/92	2·00
A	merus	A	23	20	16	41	15	53/283	62·6/283	0·85
A	merus	B	12	10	15	38	28	41/149	41·2/149	0·99
B	merus	A	17	17	16	31	12	138/228	46·6/228	2·96
B	merus	B	9	17	12	34	33	70/101	31·9/101	2·19
melas	A	A	5	12	13	14	6	71/86	20·3/86	3·49
merus	A	A	11	14	23	14	10	139/190	51·2/190	2·71

duce males with the highest degree of atrophy of the reproductive system including the accessory gland. Bryan (1972) has shown that females mated with these males will accept normal males and will subsequently lay fertile eggs.

Another outcome of this series of laboratory cage competition experiments has been the observation of higher mortality among females in those cages containing sterile males than in the control cages (in twelve of the fourteen combinations used, to be exact). This was shown in separate experiments not to be merely an effect of density increase and one is driven to wonder what happens in the field where mass releases of sterile males are being made on a daily basis against a rapidly declining natural population if there is little difference in the expectation of life of male and female. Perhaps the almost complete disappearance of females in Pala towards the end of the field trial (yet when no increase in sterile layings was apparent) could have been due in part to the sheer pressure and aggression of an accummulated, overwhelmingly high density of sterile males. Cuellar (1973b) points out that the rate of decline in numbers of normal males (proved by dissection) per room in the release village of Pala in the Upper Volta field trial was much the same as that of females in the control village while the rate of decline of females in Pala was much greater than both.

Male mortality on the other hand proved to be higher in the control cages of the laboratory experiments in ten of the fourteen combinations used in spite of the lower densities in these cages. This is a reflection of the presence of heterotic sterile males in the experimental cages. In fact, sample dissections from experimental cages showing exceptionally high mortalities demonstrated in all but the crosses involving the *An. merus* female that the sterile male survived better than the normal. In twelve of the fourteen combinations given in Table VI a comparison was made of the efficiency of young (less than 24 h old at the beginning of the experiment) and older sterile males. Four showed the young males to be more efficient than the older ones while in eight the older were better. However, some of the older males had been left with their females before being transferred to the experimental cages and evidence was obtained which indicated that these were less efficient than males of the same age separated from their females shortly after emergence.

In the original work on the crossing characteristics of the then known five species of the *An. gambiae* complex (Davidson, 1964) gross sex distortion in favour of males was observed to result from crosses between males of species A and species B and females of *An. melas, An. merus* and species C, though it was remarked at the time that on occasion normal sex ratios sometimes resulted from crosses between species A and B males and *An. melas* and *An. merus* females and since have been noted

TABLE VII Expected female (hybrid and target) fertilizations assuming random mating when different percentages of hybrid females are present.

Percentage of hybrid females in the F_1 generation	Effective ratio of sterile to normal males	Proportion of total females expected to be fertilized	Proportion of target females expected to be fertilized
Intended ratio of sterile to normal males 2:1			
0	2:1	0·33	0·33
10	1·8:1	0·36	0·30
20	1·6:1	0·38	0·28
30	1·4:1	0·42	0·26
40	1·2:1	0·45	0·25
50	1:1	0·50	0·25
Intended ratio of sterile to normal males 9:1			
0	9:1	0·10	0·1
10	8·1:1	0·11	0·06
20	7·2:1	0·12	0·04
30	6·3:1	0·14	0·04
40	5·4:1	0·16	0·03
50	4·5:1	0·18	0·03

where species C females were involved as well (no obvious sex distortion has yet been detected in any of the crosses involving species D). The reason for this sporadic "breakdown" of the sex distortion mechanism is not understood. A number of attempts have been made over the years to "perfect" it by selection from individual familes of species A from Pala and of *An. melas* but without success. The average proportion of females appearing in the F_1 generation of the cross is one-third. Some theoretical consideration has been given to the effect of the appearance of varying proportions of hybrid females in the F_1 generations of those crosses expected to give all male offspring and the release of these females along with sterile males. The general effect is to lower the ratio of sterile to normal males and thus increase the proportion of females receiving normal spermatozoa, though some of these females might be hybrids. Assuming random mating and that every female becomes fertilized in the control, then at a 2:1 ratio of sterile to normal males the proportion of all females expected to receive normal spermatozoa would rise from 0.33 to 0·5 with a rise in the percentage of hybrid females from 0 to 50 but the proportion of normal females receiving normal spermatozoa would fall from one-third to one-quarter. Equivalent figures for a 9:1 ratio of sterile to normal males are 0·10 to 0·18 and one-tenth to one-thirtieth respectively (Table VII).

If the F_1 generation of a cross between two species is being used against a third species then any mating which takes place between a wild male and a hybrid female will produce sterile male and fertile female offspring ("triple hybrid"). If it is being used against one of the two species of the cross then a proportion of the male offspring of the cross between normal male and hybrid female will be fertile and a continued backcrossing of this type will lead to a rise in this proportion of fertile males. If, however, the sterile male prefers to mate with the hybrid female and the normal male will not mate with the hybrid female (which appeared to be the case in the Upper Volta field trial) then that many less sterile male matings would occur with normal females and the situation would be equivalent to reducing the proportion of sterile to normal males, namely to lessen the overall sterilizing effect. This effect is not very marked up to the level of 30% of the F_1 generation being hybrid female and certainly not so when the higher ratio of sterile to normal males is concerned (Table VIII).

TABLE VIII Expected target female fertilizations when different percentages of hybrid females are present and the hybrid female only mates with the sterile male and the latter will not then mate with the target female.

Percentage of hybrid females in the F_1 generation	Effective ratio of sterile to normal males	Proportion of target females expected to be fertilized
Intended ratio of sterile to normal males 2:1		
0	2:1	0·33
10	1·6:1	0·39
20	1·2:1	0·45
30	0·8:1	0·55
40	0·4:1	0·73
50	0:1	1·00
Intended ratio of sterile to normal males 9:1		
0	9:1	0·10
10	7·2:1	0·12
20	5·4:1	0·16
30	3·6:1	0·22
40	1·8:1	0·36
50	0:1	1·00

(v) *Summary*

Summarily then it can be stated that in the *An. gambiae* complex of six sibling species we have the potential of employing the sterile hybrid males (produced by crossing any two of these species, in one direction at least) as a measure of genetic control of natural populations. The efficiency of these sterile hybrid males has been amply demonstrated in the laboratory in the confines of cages. In addition to hybrid sterility, sex-ratio distortion resulting in a predominance of males exists in some of the crosses. This, along with the relative ease of mass-rearing of most of the species, makes for mass-production efficiency. A single field trial has been made to date. This involved the release of the sterile hybrid males (and a few fertile hybrid females) from a cross between species B male and *An. melas* female into a declining, isolated, natural population of species A in Upper Volta. Ovipositions from matings between introduced sterile male and the natural female were few and several explanations have been offered for this dearth. Subsequent laboratory observations have indicated that mating barriers may be of some importance but not of overriding significance. There is in addition some laboratory and field evidence of an aggressive effect of the sterile males on female mortality.

The suggestion has been made that before a second field trial is carried out that large field-cage trials on the lines of those used for other insects should be considered. The usefulness of such an approach is entirely dependent on conditions being realized which allow the natural mating of wild populations. These are easy to produce in the case of the mosquitoes *C. p. fatigans* and *Ae. aegypti* which, in any case, will mate in the smallest of cages. As far as is known no-one has ever succeeded in producing artificial conditions which allow the natural swarming and mating of anophelines. A series of cages of different sizes would have to be constructed and seeded with field-collected pupae or at least the F_1 generation of wild-caught females. That size of cage (and this might be different for each species) which produced a significant insemination in a reasonably short period of time (about one week) could be used to test the competitiveness of hybrid males produced first of all from mass-matings of established colonies. If this produces a significant sterilizing effect and/or effect on female mortality in a reasonable sterile to normal ratio then little more need be done in the way of large-cage experiments. If it does not then it would be necessary to adopt the more cumbersome procedure of crossing wild populations using the artificial mating technique (as has already been emphasized, wild populations of different species rarely hybridize in nature). If this works then the failure of the previous field trial through lack of mating is proved though the future of the technique as a practical means of genetic control is uncertain, to say the least. If it does not work then the whole concept of hybrid sterility for genetic control is in jeopardy.

C. The Anopheles punctulatus Complex

Another anopheline complex currently under investigation is the *Anopheles punctulatus* complex of New Guinea and the surrounding area. Originally three species were recognized on morphological features, viz., *An. punctulatus*, *An. farauti* and *An. koliensis*. Laboratory crossings have now shown *An. farauti* from Northern Australia to produce a sterile hybrid generation when mated with *An. farauti* from New Britain and thus four species are recognized (Bryan and Coluzzi, 1971; Bryan, 1973a). The twelve crosses between these four species all produce an F_1 generation but in three death occurs in the larval stage. In the other nine, both the male and the female hybrid adults are sterile (Bryan, 1973b). The sterile males and sterile females produced by crossing the two species of *An. farauti* (provisionally called *An. farauti* No. 1 and *An. farauti* No. 2) have been used in cage competition experiments on similar lines to those made with *An. gambiae* species hybrids. Reciprocal hybrids were tested against the New Britain population of *An. farauti* No. 1 as this was the only population showing high cage fertilizations. 1:1 and 2:1 ratios were investigated. The sterile males proved uncompetitive and deliberate consecutive matings showed the accessory gland to be inactive. The sterile females, on the other hand, successfully competed with the normal females for the normal males and their addition significantly decreased the fertilization rate of the normal females (Bryan, 1973c). However, as the author points out, such females could hardly be used in the field unless they had been selected for animal, rather than man-feeding.

D. Other Anopheline Crosses

The only other cross between anopheline species known to produce a viable adult hybrid generation is that between *An. litoralis* of the Philippines and *An. sundaicus* of the Indo-Malaysian area. These are both coastal salt-water breeders with small morphological differences. Reciprocal crosses have been made and result in hybrid male sterility and hybrid female fertility. No competition experiments have ever been made with the sterile males. Coluzzi *et al.*, (1971) after studying the chromosomes of *An. stephensi* from Iraq, Iran, Pakistan and India and *An. superpictus* from Italy and noting many similarities succeeded in crossing them both ways (one involving the artificial mating technique). When the parent female was *An. stephensi* the F_1 generation died in the first larval instar. When the parent female was *An. superpictus* death occurred later, but before pupation.

Several other anopheline complexes have been described on morphological grounds, e.g., *An. hyrcanus*, *An. barbirostris*, *An. umbrosus*, *An. funestus*, *An. leucosphyrus*, *An. claviger*, etc., but no crosses have yet been made between the member "species". On the other hand numerous crosses have been made between geographical populations of the same species, mainly in studies on the mode of inheritance of insecticide resistance:

An. stephensi from Iraq, Iran, India and Pakistan (Davidson and Jackson, 1961; de Zuleta *et al.*, 1968).

An. albimanus from Panama, El Salvador, Haiti and Venezuela (Davidson, 1963a).

An. quadrimaculatus from Maryland, South Carolina and Florida (Davidson, 1963b).

An. sundaicus from Java and Malaya (Davidson, 1957; Soerono *et al.*, 1965).

An. pseudopunctipennis from the Pacific and Atlantic coasts of Mexico (Martinez-Palacios and Davidson, 1967).

An. pharoensis from Egypt and Nigeria.

An. sacharovi from Greece, Turkey and Syria.

An. f. funestus from Ghana, Upper Volta, Nigeria, Kenya and Ethiopia.

No evidence of hybrid sterility was found in any of these crosses.

II The *Aedes mariae* Complex

Among culicine mosquitoes there is evidence of a complex in *Aedes mariae*, a salt-water breeding species occurring on the coasts of various Mediterranean countries. Three species are recognized: *Ae. mariae* from the western Mediterranean, *Ae. zammitii* from the Adriatic coast of Italy and the coasts of Greece and Turkey and *Ae. phoeniciae* from Cyprus, Lebanon and Israel. Crosses between *Ae. mariae* and *Ae. zammitii* produce sterile males, in which spermatogenesis is evident but where the spermatozoa are abnormal, and fertile females (Coluzzi and Sabatini, 1968b). The other crosses produce sterile males and at least partially-sterile females (Coluzzi *et al.*, 1970).

III Tsetse Fly Crosses

Several crosses have been made between species of tsetse-flies (Vanderplank, 1948) and an attempt to control *Glossina swynnertoni* by deliberately releasing *G. m. morsitans* was described by Vanderplank (1947). Pupae of the latter species collected in the wild were placed in an area where an isolated population of the former species existed. The *G. m. morsitans* lived long enough to successfully hybridize with the *G. swynnertoni* but the environment favoured by the latter was unsuitable for the establishment of the former. The net result was a rapid decline in the *G. swynnertoni* population followed by a decline in the numbers of *G. m. morsitans* but no long-term follow-up observations were made. More recently Curtis

(1972) describes crosses between two populations of *G. m. morsitans* from Rhodesia and Tanzania and populations of *G. m. centralis* and *G. m. morsitans ugandensis*. Only a few viable offspring resulted when the crosses involved the males of *G. m. morsitans*. More were derived from reciprocal crosses but the F_1 males were fully sterile and F_1 females partially so. It has been suggested that *G. m. morsitans* males might act as effective "sterile males" for the control of either of the other two subspecies though whether there would be barriers to mating in the wild is unknown.

IV *Teleogryllus* Crosses

The use of hybrid sterility to introduce a non-diapausing trait into a diapausing species has been described by Hogan (1971). The diapausing pest species is the cricket *Teleogryllus commodus* which infests pasture land in southern Victoria, Australia. A closely related non-diapausing species *T. oceanicus* exists in northern Queensland. The two species differ slightly in egg-morphology and in the size and spacing of the stridulatory teeth of the tegmina. They readily hybridize in the laboratory and the F_1 generations are mostly sterile particularly when *T. oceanicus* male is crossed with *T. commodus* female. This cross also invariably produces F_1 eggs which are non-diapausing. Male field releases at 5:1 and 10:1 ratios of non-diapausing to diapausing populations produced 51% and 74% cross fertilizations instead of the expected 83% and 91% though 94% of all the recovered females laid at least one batch of non-diapausing eggs, indicating that cross-mating with the non-diapausing population had occurred at least once with each of these females. By crossing and back-crossing the two species it is hoped to produce a non-diapausing population with stridulating characteristics approaching those of *T. commodus* and then the cross-mating proportions should improve.

V Reduviid Bug Crosses

Hybrid sterility has also been demonstrated in reduviid bugs of the genera *Triatoma* and *Rhodnius* in Brazil (Perlowagora-Szumlewicz and Correia, 1972). Six species have been investigated *T. infestans*, *T. sordida*, *T. brasiliensis*, *T. maculata*, *R. neglectus* and *R. prolixus*. Thirteen crosses were attempted but only five produced an F_1 generation. In all five the F_1 male was sterile and the F_1 female partly fertile. Hybrid vigour was evident and male dissections showed some mature spermatozoa. When such hybrid males mated with parent females spermatophores were transferred but the female spermathecae remained empty. Whether such matings prevent subsequent normal matings remains to be determined.

6. Cytoplasmic Incompatibility

To properly understand the term cytoplasmic incompatibility, it is necessary to know the sequence of events which precede and lead up to the fertilization of the egg in mosquitoes, where this phenomenon is best known. Pairing of male and female is followed by the act of copulation, which leads to insemination of the female or the passage of spermatozoa into special sperm-storage organs, the spermathecae, where they remain viable for most of the female's subsequent life, and are utilized to fertilize successive batches of eggs. Fertilization of the eggs only occurs when they pass down the oviduct prior to actual oviposition. As each ovum passes the spermathecal duct, one or more spermatozoa enter through the micropyle. Entry triggers off the two meiotic divisions of the egg nucleus, and it is only when these divisions are completed, which may be as long as 30 min. after oviposition, that a single haploid spermatozoan nucleus migrates through the cytoplasm of the egg to fuse with one of the four haploid nuclei, the pronucleus, of the ovum. Then mitotic divisions of the new diploid nucleus proceed to produce the embryo of the new individual mosquito.

The situation where a cross between two populations of apparently the same species results in insemination without fertilization though with partial embryonation in some ova has been termed cytoplasmic incompatibility, from the fact that spermatozoa are known to enter the cytoplasm of the egg, but no actual fusion between spermatozoan nucleus and ovum nucleus occurs. The phenomenon is best known in mosquitoes of the *Culex pipiens* complex.

I. The *Culex pipiens* **Complex**

Taxonomically the *C. pipiens* complex is considered to consist of three subspecies, *C. p. pipiens*, *C. p. fatigans*, and *C. p. australicus*, with in addition a varietal name *molestus* applied to autogenous populations able to mature eggs and oviposit without feeding on blood. While *C. p. pipiens* is more characteristic of the temperate regions of the world, and *C. p. fatigans*, the filariasis vector, is usually found in tropical and subtropical areas, there is a broad belt of overlap. *C. p. australicus* is restricted to Australia, New Caledonia and the New Hebrides, and is often sympatric with *C. p. fatigans*.

Though several workers had previously reported unsuccessful attempts to cross populations of the subspecies *C. p. pipiens*, and to cross this sub-species and *C. p. fatigans*, Laven (1951) was the first to examine the sperma-thecae of females laying sterile egg rafts from such crosses and to find spermatozoa. Moreover, most of the ova were embryonated. His original crosses were the reciprocal ones between two autogenous strains of *C. p. pipiens*, one from Hamburg and one from Paris. Sterile rafts were laid from both crosses. However, when a third population from Oggelshausen in southern Germany was crossed to both of these strains, a different result was obtained from those crosses involving the male of the Oggel-shausen strain. Both crosses to females of the Paris and Hamburg strains resulted in fertile offspring, while the reciprocals again produced insemina-tion of the Oggelshausen females, which oviposited, but the egg-rafts did not hatch, even though most of the eggs were partially embryonated. This then was a uni-directional incompatibility as distinct from the bi-directional incompatibility exhibited when the Paris and Hamburg populations were crossed.

F_1 females derived from the cross between Oggelshausen male and Hamburg female are reproductively normal, and can be further crossed to Oggelshausen males to produce viable offspring. In fact, these F_2 hybrid females can be again crossed to Oggelshausen males to produce normal F_3 hybrid females, and the procedure repeated again and again. Laven (1959) carried out such repeated backcrosses through 52 genera-tions over a period of five years. The net effect of repeated crossing to the Oggelshausen strain was to gradually replace the genome (the chromo-somes) of the Hamburg strain with that of the Oggelshausen strain. In spite of this replacement, the hybrid females and males from the final cross still remained "Hamburg" in crossing-type. That is to say, the female produced viable offspring when further crossed to the Oggelshausen male, but the male remained incompatible with the Oggelshausen female. Laven concluded, therefore, that the determinants of incompatibility are extra-chromosomal or cytoplasmic. He further showed that the crossing-type characteristic is completely stable, and in a publication of 1971 states that the Hamburg and other strains first studied in 1947 still retain their original characteristics.

Laven (1967a, 1969a and b, 1971) and others, e.g., Dobrotworsky (1955), Barr (1966), Eyraud and Mouchet (1970) and Krishnamurthy (1972) went on to cross populations of all three subspecies of the *C. pipiens* complex from many different parts of the world, and now at least 20 different crossing-types are recognized. Some of the crosses produce low hatches and both male and female offspring. These have since been termed "partially compatible" by Barr (1969), who found difficulty in distinguishing some of them from totally incompatible crosses. Laven (1957) did not recognize crosses producing both male and female offspring as incompatible but only those showing something of the order of 0·1 to 0·2% hatch and producing only female offspring. Jost (1970a and b) has made a detailed study of embryogenesis in incompatible crosses, and finds that though up to 75% of the eggs may show development even to the extent of body segmentation and hair and eye development, 99·9% die within the egg-shell and only about 0·1% hatch to produce fertile diploid females, compatible with the maternal parent population. Such development is considered meiotic parthenogenesis stimulated by penetration of the ovum by the spermatozoon. Dying embryos are in fact haploid, while in those few surviving females diploidy is restored by a change in the meiotic process in the egg (1970b), or by fusion of the pronucleus and the last polar body (1970a).

There has been considerable speculation as to the nature of the cytoplasmic determinants of incompatibility. Laven (1967a, 1972) favours the existence of plasmagenes, and dismisses the idea of a virus, or some such pathogen transovarially transmitted, on the grounds that a number of such agents equivalent to the number of crossing types must be postulated. Evidence of such agents has now been demonstrated by Yen and Barr (1971) in the form of *Rickettsia*-like bodies in the egg and concentrated near to the micropyle, in both normal and incompatible crosses within the subspecies *C. p. pipiens*. It is suggested that some twenty strains of this micro-organism exist and are responsible for the "blocking" of spermatozoa in incompatible crosses, preventing actual fertilization. McClelland (1967), on the other hand, favoured the idea of cytoplasmic conditioning genes on chromosomes, from suggestions put forward by Smith–White and Woodhill (1955), to explain incompatibility among crosses in another mosquito complex (*Aedes scutellaris*).

Without going into further detailed discussion on the actual causes of incompatibility, the fact remains that a number of crosses in the *C. pipiens* complex produce no offspring or very, very few. The latter are invariably female, and of the same crossing type as their mother. Thus the possibility exists of releasing an incompatible strain to eliminate wild populations. The release material must of course be only of one sex. Otherwise the release population could inter-mate and produce its own offspring. The sex of choice must be male, as this sex does not bite and transmit disease, and because it is capable of mating more than once. The very few females produced from the incompatible cross would be incompatible with release males. According to Laven (1971) there is no question of any mating barrier between release and wild material as both belong to the same species and differ only cytoplasmically. No difficulties were in fact met with in this respect in cage experiments. However, when it came to transferring attention to the control of field populations in the tropics using incompatible strains from temperate climates, it became necessary to incorporate some of the genetic background of tropical populations by appropriate backcrosses. As pointed out already, this change of genome in no way affected the compatibility status.

For a first field trial a small village near Rangoon in Burma was chosen (Laven, 1971). This was the village of Okpo, 25 km north-west of the capital city. In the dry season the mosquito population, breeding mostly in artificial containers, e.g., clay pots and metal containers of various forms in and around the houses, was considered free from invasion from breeding sources in nearby villages. In the village were some 700 people occupying 145 houses.

The local population of *C. p. fatigans* was colonized in Germany and tested for incompatibility with nine strains of *C. p. pipiens* or *C. p. fatigans* from Germany, France, Italy, Egypt, U.S.A. and Japan. It proved bidirectionally incompatible with three of them (from Oggelshausen, Paris and Kawasaki, Japan) and unidirectionally incompatible with two (from Hamburg and Fresno, California). The Paris strain was chosen for a start in the production of an incompatible strain, showing as it did a superior fitness in most aspects of reproduction. However, it was of the subspecies *C. p. pipiens* and adapted to a temperate climate. To introduce a tropical background it was crossed and backcrossed through eight generations, either with a strain from Fresno, California or with one from Freetown, Sierra Leone. Laven (1971) admits doubt as to which strain was actually used, as there was some evidence of prior mixing or substitution in his laboratory. Be that as it may, a reconstituted strain called D1 was evolved and transferred to Burma. There it was crossed in cages to 25 samples of the local *C. p. fatigans* collected from more or less equidistant points in a around the village of Okpo. Rangoon being a major port would be continually subject to importation of foreign strains of the *C. pipiens* complex. Hence it was imperative to eliminate any chance of compatibility between the D1 strain and local populations. Thus 25 crosses between local females and D1 males were set-up in cages. From these, 880 egg-rafts comprising more than 130 000 eggs were counted and kept to see if any hatched. 180 larvae (0·14% of the eggs counted with a variation of from 0–0·43%) hatched. All the emerging adults proved to be female, and nearly 72% of non-hatching eggs proved to be embryonated. Thus all 25 samples were incompatible with the D1 strain.

The mating ability of the males of the D1 strain was compared with that of local males by caging equal numbers (1 020 of each) of the same age with 1 025 local females. Both local males and local females were derived from wild-caught pupae. Of 700 egg-rafts subsequently laid by the females, 67% showed the abnormal hatches characteristic of incompatibility, and the D1 males thus appeared to be twice as efficient as the local ones in mating with the females under these conditions.

Field releases of D1 males were planned to coincide with the end of the peak production of the local mosquito population in February 1967. At that time, or shortly before, various methods of estimation of population size, including the sampling of indoor resting adult mosquitoes, the regular counting of egg-rafts in specified catching stations, and a mark-release-recapture study by Macdonald et al., (1968), suggested an emergence of 3 000 to 4 000 adults (1 500 to 2 000 males) each day. Releases of D1 males started in mid-February, but barely surpassed 2 000 per day until mid-March. During this time there had been an unexpected build-up in the wild population to a peak of something like 15 000 adults per day (7 500 males). From mid-March until the end of the first week in May a release of 5 000 D1 males per day was maintained for the most part. In all, during 80 days, 275 000 males were released with an average calculated ratio of incompatible to wild males of 1·2:1 [(Laven, 1972). The first sterile rafts began to appear less than one week after releases started, and by the end of March formed nearly one-quarter of the rafts collected (284 on March 31). By the end of April this proportion was nearly three-quarters (of 290 rafts collected on April 30), and by the ninth of May every raft collected (75) failed to hatch. The situation on May 10 was the same; of 65 rafts collected none hatched. The following day the monsoons started, and though as Laird (1967) has pointed out, the total eradication claimed by Laven (1967b) is not correct, as some rafts had still hatched in days previous to May ninth, and aquatic stages were still in existence, the efficiency of the method is obvious, and if carried out in a really isolated population (which this certainly was not in the rainy season), eradication could surely have been achieved by continuing releases for a further two weeks or so (the length of the aquatic cycle plus the time to maturity and receptivity to mating of the adult female) after 100% sterility of the rafts was first achieved.

A second field trial did not take place until 1972, when a comparison was made of this method with the release of irradiated and chemosterilized C. p. fatigans males within the intended programme of the World Health Organization/Indian Council of Medical Research Unit for the Genetic Control of Mosquitoes in Delhi, already referred to. First a strain bidirectionally incompatible with four Indian populations from Ahmedabad, Bangalore, Delhi and Madras was constructed.[1] This D3 strain has the

[1] Krishnamurthy, B. S. and Laven, H. (1972). World Health Organization mimeographed document WHO/VBC/72. 389.

cytoplasm of the Paris strain and the genome of the Freetown strain. Later a D4 strain, bi-directionally incompatible with the D3 strain and uni-directionally incompatible with the Delhi strain (D4 male is compatible with Delhi female), was made utilizing the cytoplasm of the Hamburg strain and a combined genome from alternative backcrossings to populations from Bangkok and Dar-es-Salaam. This "back-up strain", as it was called, was produced as a safeguard to meet the possible eventuality of some D3 females being released along with D3 males, and the foreign strain establishing itself in the wild.

The D3 males proved comparable with Delhi males in mating competitiveness and survival in small cage experiments conducted both in the laboratory and outside in Delhi.[1] A field release of some 7,000 marked males of each of the D3 and Delhi populations showed equal rates and distances of dispersal over the following six days, though the survival rate of the D3 strain proved to be significantly lower than that of the local strain. Laboratory observations indicated differences in oviposition time, pupal period and male mating age between the two strains. The D3 strain tended to oviposit in the early morning or after daybreak, had a longer pupal period, and the male was not at its mating peak until some 12–24 h after that of the Delhi male.

In March 1972 a direct comparison of the efficiency of D3 and thiotepa-sterilized males was made in two villages near Delhi. The pre-release populations were estimated to be in the region of 10 000–30 000 per village with some 1 500–5 000 males emerging each day. Until mid-May releases were of the order of 10 000 males per day, but in both villages egg-raft sterility remained low—10% in the village where the chemosterilized males were released as compared with 4% in the village where the incompatible males were released. Later, however, when the chemosterilized male release was increased to 100 000 per day, 75% sterility was achieved (in the second week in June). The incompatible strain proved more difficult to mass-rear, and the removal of every single female before release took time and manpower. It was actually done in a room cold enough to immobilize the mosquitoes, and took five man-hours to process 10 000 specimens as compared with a single man-hour to sterilize 100 000 pupae with thiotepa. In spite of these difficulties releases of D3 males were increased to near 23 000 per day at the end of May and the beginning of June and kept to more than 12 000 per day until June 10. Egg-raft sterility in this last week of release increased to 27% and in the following week to 62%.[2] After this trial a general opinion was expressed that because of the imperative necessity of 100% sex-separation in the incompatibility tech-

---

[1] World Health Organization mimeographed document (1972)—report of a technical planning and review group—VBC/72.2.

[2] Information derived from monthly reports of WHO/ICMR Research Unit on the Genetic Control of Mosquitoes, Delhi.

nique, that this could not be considered operationally feasible for large-scale work.

An alternative use of the D3 strain was then suggested. If able to reproduce itself in the wild and over-released as both sexes, it might then replace the existing population. If it could be selected for resistance to filarial infection, prior to its release, then the existing vectorial population might be replaced by a refractory one. The two strains, D3 and Delhi, were in fact fed on human microfilarial carriers, and both proved highly susceptible to infection. However, the possibility of future selection (for refractoriness) remained, and experiments were made to compare the fitness of the two strains by releasing them together in large, outdoor, walk-in-type cages in the Delhi winter conditions (December 1972–March 1973). The outcome of these experiments was that even when the D3 strain was present to the extent of 81% of the mixed population, it declined in frequency rapidly, and was poorly adapted to local conditions.[1]

Now the D3 strain has been combined with a male-linked translocation complex 71 with a Paris genome and conferring a sterility of 85% ± 10 to its own females or to other populations of females.[2] This D3/71 strain remains bi-directionally incompatible with the local Delhi population, but is of apparently inferior fitness, because of its European genome. Experiments with this integrated strain will be described in the later chapter on translocations.

II. The *Aedes scutellaris* Complex

Apart from uni-directional incompatibility in *Mormoniella* (a parasitic wasp), and probably in *Clunio (Chironomidae)* (Laven, 1967a), the only other recorded case appears to be in another culicine mosquito complex, the *Ae. scutellaris* complex. Under this umbrella-name, as many as 23 different specific, subspecific or race names have been included, identified mainly on small morphological differences, and on geographical distribution (World Health Organization, 1964). Members of this complex are vectors of filariasis. They occur on many of the islands of Pacific Oceania. Many of the so-called species are unique to particular islands, and are considered by some to be recently evolved, or still-evolving, species. Attempts to cross some of them have demonstrated that mating barriers are not complete—Woodhill, 1949a,b and 1950, using *Ae. s. scutellaris* and *Ae. s. katherinensis*; Perry, 1950, using *Ae. hebrideus* and *Ae. pernotatus*; Rozeboom and Gilford, 1954 and Woodhill, 1954 using *Ae. pseudoscutellaris* and *Ae. polynesiensis*.

[1] Information derived from monthly reports of the WHO/ICMR Research Unit on the Genetic Control of Mosquitoes, Dehli.
[2] World Health Organization mimeographed document (1972)—report of a technical planning and review group—VBC/72.2

Woodhill (1949a,b) crossed *Ae. s. scutellaris* from New Guinea with *Ae. s. katherinensis* from northern Australia. The female of the former crossed to the male of the latter produced fertile offspring, while the reciprocal produced no offspring at all, even though spermatozoa were found in the spermathecae of the *Ae. s. katherinensis* female. Further, Woodhill (1950) found that the F_1 generation behaved like its maternal parent, and the males remained "incompatible" with female *Ae. s. katherinensis*. In fact repeated backcrosses, similar to those done by Laven (1959), failed to change the crossing characteristic (Smith-White and Woodhill, 1955). It is not inconceivable, therefore, that mass-reared *Ae. s. scutellaris* males could be used to eradicate *Ae. s. katherinensis* under appropriate conditions.

7. Translocations

I. General Considerations

A chromosome translocation involves the breakage of two non-homologous chromosomes, and the re-attachment of the broken parts to the wrong partners (sequences (a), (b) and (c) in Fig. 14). This may occur naturally, but can be produced artificially by exposure to irradiation or to radiomimetic chemicals. At meiosis in such a translocation heterozygote, the pairing of like parts of the translocated chromosomes produces a cross-like configuration which can be readily seen in cytological preparations made from germinal tissues. At nuclear division in such a cell three directions of segregation (alternate, adjacent-1 and adjacent-2) can occur to produce a total of six different kinds of gametes (sequences (d) and (e) in Fig. 14). Only two of these (combinations 1 + 4 and 2 + 3) in sequence (e) are balanced (orthoploid) in that they have the full gene complement of both chromosomes, albeit rearranged in one of them, viz., 2 + 3. The other four are unbalanced (aneuploid) in that they have duplications of the genes of one chromosome and shortages of the genes of the other (combinations 1 + 2, 3 + 4, 1 + 3 and 2 + 4). The six different kinds of gametes are not produced in equal proportions. In *C. pipiens*, for example, alternate segregation occurs more frequently than adjacent-1 or adjacent-2 (Jost and Laven, 1971). The usual net result is that 50% are balanced and viable and 50% are unbalanced and lethal. In other words a mating between a simple translocation heterozygote and a wild-type individual usually results in an approximate halving of fertility.

In many respects a translocation can be considered a simple Mendelian inheritance mechanism though with some departure from the expected 1:2:1 ratio when two heterozygotes mate:

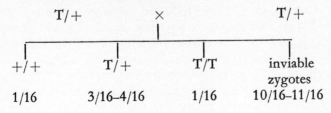

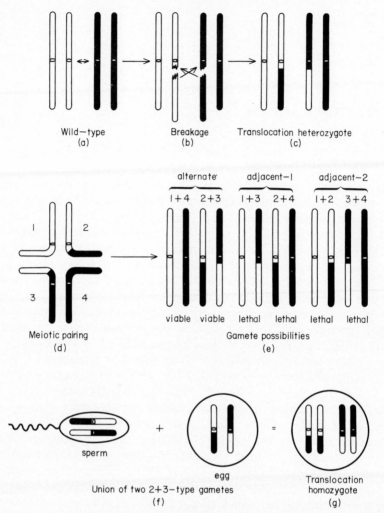

Fig. 14   Simple translocation formation.

This is because of variations in the production of the four different types of aneuploid gametes, and because a union of two complementary aneuploid gametes can produce a viable balanced heterozygote, e.g. $1 + 2$ with $1 + 3$ or $2 + 4$ with $3 + 4$ in sequence (e) of Fig. 14. From such a mating then, 16·7 to 20% of the offspring would be expected to be homozygous for the translocation (sequences (f) and (g) in Fig. 14). Whether they will be viable or not depends on how much damage has been sustained in the areas of the chromosomes surrounding the breakpoints, and what effect the changes in the position of whole blocks of genes in relation to other genes has. According to Burnham (1962) nearly one-half of 53 different translocations isolated in *Drosophila* showed full viability when made homozygous. As we shall see, however, when we come to consider isolations in specific insect pests, it would seem that viability in the homozygous state is by no means as common as this. Nevertheless, if viable, they should be fully fertile as all the resulting gametes should be orthoploid, i.e., $T/T \times T/T$ should give all T/T offspring. If mated with the wild type, however, the offspring will be all translocation heterozygotes, i.e., $T/T \times +/+ = T/+$, while matings between homozygote and heterozygote will produce 50% inviable zygotes, and half the viable progeny will be heterozygous, and half homozygous, i.e., $T/T \times T/+ = T/T + T +$.

What has been considered so far is a translocation involving two autosomes. Two other main types can occur: between an autosome and the Y-chromosome (or where sex chromosomes are not differentiated, between that part of the chromosome bearing the genetic factor determining maleness; in *Culex* and *Aedes* mosquitoes this is the dominant allele (M) of a pair and the male is heterozygous (Mm)), or between an autosome and the X-chromosome (or in *Culex* and *Aedes*, that part of the chromosome bearing the recessive sex-determining allele (m)). In the first type semi-sterility is inherited only through the male so long as no crossing-over occurs, and can be depicted as follows:

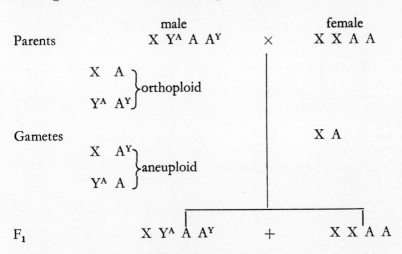

In such a population the male imparts some 50% sterility whether it mates with its own female or a wild female. In the second type, it can be inherited through both male and female but only the female can become homozygous; the male is hemizygous. A homozygous population can be depicted as follows:

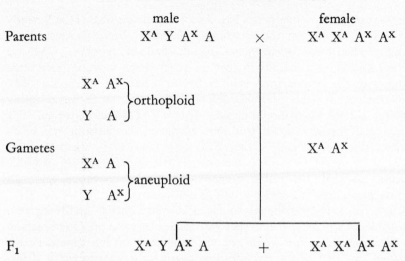

As can be seen, because half the male gametes are aneuploid, such a population will only produce half the offspring of a normal population. When outcrossed to a wild population the hemizygous male mating with the wild female will be:

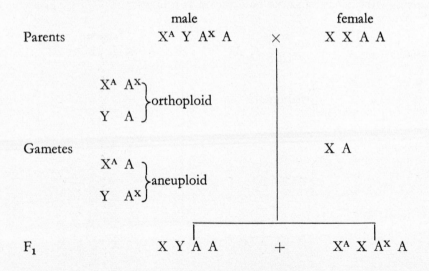

and result in a half-hatch and the production of normal males and heterozygous females. Homozygous females outcrossed to wild-type males will produce a full complement of offspring half of which will be hemizygous males and half heterozygous females:

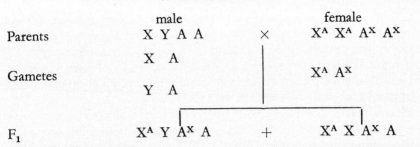

$$\text{Parents} \quad \overset{\text{male}}{X\ Y\ A\ A} \quad \times \quad \overset{\text{female}}{X^A\ X^A\ A^X\ A^X}$$

$$\text{Gametes} \quad \begin{matrix} X\ A \\ Y\ A \end{matrix} \qquad\qquad X^A\ A^X$$

$$F_1 \quad X^A\ Y\ A^X\ A \quad + \quad X^A\ X\ A^X\ A$$

Other matings producing reduced hatches will be between hemizygous male and heterozygous female and between heterozygous female and wild-type male as follows:

$$\text{Parents} \quad \overset{\text{hemizygous male}}{X^A\ Y\ A^X\ A} \quad \times \quad \overset{\text{heterozygous female}}{X^A\ X\ A^X\ A}$$

Gametes

$$\left.\begin{matrix} X^A\ A^X \\ Y\ A \end{matrix}\right\} \text{orthoploid} \left\{\begin{matrix} X^A\ A^X \\ X\ A \end{matrix}\right.$$

$$\left.\begin{matrix} X^A\ A \\ Y\ A^X \end{matrix}\right\} \text{aneuploid} \left\{\begin{matrix} X^A\ A \\ X\ A^X \end{matrix}\right.$$

$$F_1 \quad X\ Y\ A\ A + X^A\ Y\ A^X\ A + X^A\ X\ A^X\ A + X^A\ X^A\ A^X\ A^X$$

$$\text{Parents} \quad \overset{\text{normal male}}{X\ Y\ A\ A} \quad \times \quad \overset{\text{heterozygous female}}{X^A\ X\ A^X\ A}$$

$$\text{orthoploid}\left\{\begin{matrix} X^A\ A^X \\ X\ A \end{matrix}\right.$$

Gametes

$$\begin{matrix} X\ A \\ Y\ A \end{matrix}$$

$$\text{aneuploid}\left\{\begin{matrix} X^A\ A \\ X\ A^X \end{matrix}\right.$$

$$F_1 \quad X\ Y\ A\ A + X^A\ Y\ A^X\ A + X\ X\ A\ A + X^A\ X\ A^X\ A$$

From the first of these, only 37·5% of the normal offspring number would be expected with one-third of the males normal and two-thirds hemizygous (one-half of these being produced by the union of two complementary aneuploids) while two-thirds of the females would be heterozygous (again one-half being produced by the union of two complementary aneuploids) and one-third homozygous. From the second, a 50% hatch would be expected producing the two sorts of males again in equal numbers and normal and heterozygous females in equal numbers.

Of course multiple translocations can occur both between the autosomes themselves and between the autosomes and sex chromosomes, involving several breakages and reattachments. The net effect is to increase the level of inherited sterility. One example will be given, and concerns a male-linked three-chromosome double translocation isolated by Akiyama (1973) in *Anopheles gambiae* species A. This was produced by irradiating a simple Y-autosomal translocation ($T^Y F5$) previously isolated by Krafsur (1972). The two steps involved are depicted in Figure 15. The possible gametes from such a translocation ($A_1 \ A_1^{YA_1} \ A_2 \ A_2^{A_1} \ XY^{A_1}$) and zygotes produced when mated with a normal female ($A_1 \ A_1 \ A_2 \ A_2 \ X \ X$) are as follows:

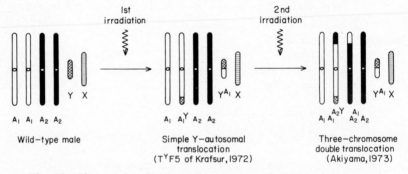

Fig. 15   The origin of a three-chromosome-double translocation.

Gametes	*Zygotes (with $A_1 \ A_2 \ X$)*	
$A_1 \ A_2 \ X$	$A_1 \ A_1 \ A_2 \ A_2 \ X \ X$	(1)
$A_1 \ A_2 \ Y^{A_1}$	$A_1 \ A_1 \ A_2 \ A_2 \ X \ Y^{A_1}$	(2)
$A_1^{YA_1} \ A_2 \ X$	$A_1^{YA_1} \ A_1 \ A_2 \ A_2 \ X \ X$	(3)
$A_1^{YA_1} \ A_2 \ Y^{A_1}$	$A_1^{YA_1} \ A_1 \ A_2 \ A_2 \ X \ Y^{A_1}$	(4)
$A_1 \ A_2^{A_1} \ X$	$A_1 \ A_1 \ A_2^{A_1} \ A_2 \ X \ X$	(5)
$A_1 \ A_2^{A_1} \ Y^{A_1}$	$A_1 \ A_1 \ A_2^{A_1} \ A_2 \ X \ Y^{A_1}$	(6)
$A_1^{YA_1} \ A_2^{A_1} \ X$	$A_1^{YA_1} \ A_1 \ A_2^{A_1} \ A_2 \ X \ X$	(7)
$A_1^{YA_1} \ A_2^{A_1} \ Y^{A_1}$	$A_1^{YA_1} \ A_1 \ A_2^{A_1} \ A_2 \ X \ Y^{A_1}$	(8)

Of the 8 different zygotes only two ((1) and (8)) would be viable, i.e., 25%.

II. Dynamics

Serebrovsky (1940) was perhaps the first person to realize the potentialities of the use of translocations to control insect populations. In fact an English translation of his paper (1969) is entitled "On the possibility of a new method for the control of insect pests". Curtis, who has done much to promote the idea since, states in his first major treatise on the subject (1968a), that his attention was only drawn to the existence of this paper after he had produced his own. Serebrovsky, then, suggested the isolation of a homozygous autosomal translocation breeding stock and the release of individuals from it into wild populations. Matings between released and wild individuals would then produce translocation heterozygotes with their inherent reduced fertility. Matings involving such heterozygotes would result in progeny reductions. Assuming an initial introduction of a single strain with a single translocation in numbers equal to those of the wild population and random mating one would expect a 50% reduction in fertility by the F_2 generation, just over 42% in the F_3 and stabilizing to 43% in subsequent generations. Releasing a strain carrying two translocations among four pairs of chromosomes would produce an overall mortality of 67·5%, i.e., 43 + 43% of 100–43. On this line of reasoning, three translocations among six pairs of chromosomes would produce 81·5% lethality, four translocations 89·5%, five translocations 94% and six translocations 96·5%. No two translocations involving the same two pairs of chromosomes can be expected to be the same, and the effect of releasing two or more different translocation stocks in which exchanges have taken place between the same chromosomes can similarly increase lethality. As Serebrovsky points out, the effect of the release of two such stocks would be to increase the heterozygotes from 50–67% in the first generation, and this would lead to an overall mortality of 54·6% instead of 50%. Increasing the number of such stocks released would eventually achieve something of the order of 75% mortality. The corresponding limit where releases of many strains with two translocations each are made would be 93·75%, and with three translocations 98·4%. As we have already seen, more than one exchange between two non-homologous chromosomes can occur and this again is accompanied by increased lethality.

These relatively simple calculations by Serebrovsky have been considerably elaborated by Curtis and Robinson (1971), Curtis and Hill (1971), Whitten (1971a,b), and McDonald and Rai (1971). McDonald and Rai (1971), concerned with the rarity of viable translocation homozygotes, advocate the release of heterozygotes, though appreciating the difficulties of selecting out such heterozygotes from mixed stocks. Even with marker-identification, selection on a large scale would present problems. All the other authors favour the release of translocation homozygotes, and for population elimination purposes, Curtis and Robinson (1971) make out a strong case for the release of double translocation heterozygotes produced by crossing two homozygous translocation stocks. The sterility of the double translocation heterozygote will be considerably higher than that of either single heterozygote, and any weaknesses in either homozygote may be offset by hybrid vigour. This they consider within the bounds of practical possibility. The multiple translocation stocks considered by Whitten (1971a,b) are somewhat too theoretical in conception, though they make the point that the more complex the system the greater the degree of sterility introduced. This author calculates that a single release of four translocation strains in equal numbers to the wild population can cause a population reduction equivalent to a 20:1 sterile-to-normal-male release for five generations (Whitten, 1971a).

Serebrovsky was appreciative of the fact that the incomplete mortalities inflicted on a wild population by the introduction of translocations, even though quite high, need not necessarily affect the ultimate population size. A reduction in the number of new individuals born in the population can be compensated for by their increased likelihood of survival in an environment where there is less competition for food and space. With insects with a high reproductive potential, in the most favourable breeding season, this compensatory influence is likely to be at its highest, and quite high mortalities are unlikely to have much effect. With insects of low reproductive potential, and even with those of high reproductive potential in unfavourable seasons, an incomplete mortality may just tip the scales from survival at a threshold level to eradication. What the particular mortality necessary to achieve elimination is depends on the species and the season, and for most species is yet to be determined.

Serebrovsky was also concerned about the instability of the equilibrium finally attained after the incorporation of translocations into wild populations, and clearly saw the possibilities of elimination of the translocations. If they were less fit or in too low a number, or of the opposite effect, namely the replacement of the wild population by the translocation ones or ones, if they were more fit or in too high a number. What Serebrovsky saw as a weakness, however, Curtis (1968b) and later Whitten (1971a,b) saw as a distinct advantage. By exceeding the proportion required to maintain equilibrium, the translocation population could act as a population replacement mechanism, and be used to introduce a harmless insect to fill the ecological niche of a harmful one, if attributes favourable to man could be "built in" to the translocation population. Such attributes could be genes controlling susceptibility to insecticides, refractoriness to human, animal or plant diseases or conditional lethal genes.

Whitten (1970) advocates the replacement of insect populations which have become resistant through the prolonged use of a particular economic insecticide by susceptible populations, and the continued use of the insecticide instead of change to another one, almost certainly more expensive and less efficient. The simple genetic nature of many of the insecticide resistances is well known, and the incorporation of susceptible alleles into translocation populations should present no difficulties.

The mode of inheritance of the ability to transmit disease is less well known, though the few cases worked out in detail show indications of simple mechanisms. As long ago as 1931, Huff obtained evidence of a genetic basis for the susceptibility of *Culex pipiens* to the chicken malaria parasite (*Plasmodium cathemerium*), and more recently Dennhofer (1971) has shown that this susceptibility is dependent on a single recessive gene. Ward (1963) found evidence of a single, incompletely dominant factor governing the susceptibility of *Aedes aegypti* to *P. gallinaceum,* another chicken malaria parasite, while Kilama and Craig (1969), using the same model came to the conclusion that a single recessive gene is involved. It remains to be seen whether the susceptibility of anopheline mosquitoes to human malaria is simply inherited or not.

The ability of *Ae. aegypti* to support infections of the subperiodic form of the filarial worm *Brugia malayi* is also simply inherited. A single sex-linked, recessive (f^m) 3·4 units from the sex locus (m) is responsible according to Macdonald (1967), and has the pleiotropic effect of supporting infections with the related periodic *B. malayi, B. pahangi* and periodic and sub-periodic *Wuchereria bancrofti,* though not *Dirofilaria immitis* or *D. repens.* An independent sex-linked recessive has recently been found to control susceptibility to infection with *D. immitis* in the same mosquito species[1]. A single sex-linked recessive has also been found to control the susceptibility of *C. pipiens* (? *fatigans*) to *B. pahangi*[2].

Among agricultural pests Macdonald (1967) cites a single sex-linked dominant gene as responsible for the susceptibility of the leafhopper (*Cicadilina mbila*) to a virus causing streak-disease in maize and the possibility of an incompletely-dominant sex-linked gene controlling the susceptibility of another leafhopper (*Aceratagallia sanguinolenta*) to potato yellow-dwarf virus.

Against such favourable evidence of the existence of simple genetic systems controlling the ability to transmit disease agents, is the evidence that the disease agents themselves are capable of undergoing genetic changes too, and that refractory insects may become susceptible again to

[1] McGreevy, P. B. and McClelland, G. A. H. (1973). World Health Organization mimeographed document. WHO/FIL/73. 109.
[2] Vector Genetics Information Service 1972. World Health Organization mimeographed document. VBC/G/73.1.

changed pathogens. Using *An. stephensi* and *P. gallinaceum* Coradetti *et al.*, (1970) have succeeded in selecting a population of the plasmodium, capable of infecting a strain of *An. stephensi* previously refractory and selecting a population of *An. stephensi* able to support an infection which would not previously infect it, while Walliker *et al.*, (1971) have been able to actually demonstrate genetic recombination in a malaria parasite. Using two lines of the rodent malaria parasite *P. berghei yoelii,* one with an identifiable enzyme mutant, and resistant to the antimalarial drug pyrimethamine, and another with a different enzyme mutant, and susceptible to the drug, they were able to recover recombinant types from a mouse infected with both lines.

The possible use of conditional lethal genes for the control of insect populations has been considered by Smith (1971). These are genes which would allow the mass-cultivation of an insect under the controlled conditions of laboratory or factory, but which would be lethal under field conditions. High and low temperature lethals are known as well as genes preventing diapause. Thus an insect conditioned to life at a high temperature could be released some time before cold external conditions are expected, or a mutant unable to enter diapause could be released in areas where diapause is essential for survival. Ideally such lethals should be dominant.

Curtis (1968b) suggests methods of construction of such translocation replacement populations. It will be essential that the gene to be introduced will remain linked to the translocation chromosomes and not be lost through crossing-over. For this purpose he recommends the induction of an inversion to include the locus of the gene to be introduced, once this has been isolated and made homozygous. The inversion must then be made homozygous, and a translocation with one break-point within the inversion isolated. This must also be made homozygous. *Drosophila* geneticists will undoubtedly recognize the feasibility of such a construction programme, having plenty of marker genes at their disposal. In most insect pest species, however, such marker genes have still to be found, and no such fully constructed transport population has yet been realized. Moreover, as we shall see, only a very few homozygous autosomal translocations have yet been isolated, and only two or three attempts have been made to release translocations of any description into wild populations.

III. The Isolation of Translocations

It might be useful at this point to comment on the methods actually used to isolate translocations in the few insect pests where this has been attempted so far. The method usually starts with the irradiation of young males at a dose somewhat below that producing full sterility. These males are then mated to virgin females. Where no suitable markers are available, the procedure is to inbreed the F_1 generation, and to then take successive ovipositions from single females. Those females producing consistently low hatches are kept, and their offspring outcrossed to the wild type, all the time looking for partially sterile matings. This process may have to be repeated through more than one generation to concentrate semi-sterile types.

Usually male-linked translocations become recognized first, from the fact that all male outcrosses produce semi-sterility while female outcrosses remain normal. X-autosomal interchanges are indicated when the daughters of semi-sterile males outcrossed to wild-type males prove semi-sterile, and their brothers normal in fertility. Eventually the stage may be reached where most of the members (both sexes) of selected families give reduced hatches on outcrossing. When this occurs inbreeding is resorted to in the hope that heterozygote will mate with heterozygote producing translocation homozygote. Then a search must be made for females giving a normal hatch, but whose sons and daughters both produce offspring from outcrosses which generate semi-sterility on further outcrossing. This method depends entirely on partial sterility as a means of recognition of heterozygous translocation matings and, of course, natural variations in control hatches must be taken into account.

A. Anopheline Mosquitoes

Such tedious methods had to be used in the isolation of translocations in *An. gambiae* species A and B (Krafsur, 1972; Akiyama, 1973), but have not yet resulted in the isolation of a homozygous autosomal translocation. Some of the heterozygous translocations have been confirmed cytologically in these species (Hunt and Krafsur, 1972). Six were simple autosomal exchanges, while one was a two-autosome double. Cytological identification of six heterozygous translocations in *An. albimanus* has also been made by Rabbani and Kitzmiller (1972). These included autosomal-autosomal, X-autosomal and Y-autosomal translocations. Again, no homozygous autosomal translocation could be isolated.

B. *Aedes aegypti*

Where good chromosome markers exist the recognition of translocations becomes much easier. Morphological features existing singly in individuals become combined when translocations take place involving the loci of such features. As an example, Rai and McDonald (1972) describe how translocations have been isolated in the mosquito, *Ae. aegypti*, vector of dengue, haemorrhagic and yellow fevers. This mosquito, so easy to rear and maintain in the laboratory, has proved to be an excellent experimental animal, and nowhere more useful than in the study of genetics (see Craig and Hickey, 1967 and McClelland, 1967). One great advantage of it is the fact that its eggs can be stored for months on end and still remain viable. Thus mutants once isolated can be "shelved" and do not have to be continuously maintained. One disadvantage is that good readable polytene chromosomes so useful in *Drosophila* and anopheline genetics studies are difficult to produce. More than 90 mutants have been described in *Ae. aegypti*, and 30 of them have been mapped in three linkage groups. Of the three pairs of chromosomes two are metacentric and one sub-metacentric. As already explained, the specific X and Y chromosomes of anopheline mosquitoes and many other insects are not present in this species; the smallest pair of chromosomes (chromosome I) bears a single locus (m) responsible for sex determination. In the homozygous state (mm) the insect is female, while in the heterozygous state (Mm), it is male.

As Rai and McDonald (1972) describe, a multiple marker RED stock exists possessing recessive genes for *red-eye* (re) on chromosome 1, *spot-abdomen* (s) on chromosome 2 and *black tarsi* (blt) on chromosome 3. To isolate translocations young males of a wild stock, ROCK, were irradiated and mated to the RED marker stock. F_1 males from those families showing 50% or less fertility were then backcrossed to RED, and those individuals in the offspring of this backcross showing "pseudo-linkage" between marker genes (*re* with *s, re* with *blt, s* with *blt*) picked out as potential translocation stocks. The more complicated procedure adopted for the isolation of homozygotes is described by Lorimer *et al.,* (1972).

The first record of the isolation of a translocation in *Ae. aegypti* is to be found in Rai and Asman (1968), where a male-linked translocation involving chromosomes 1 and 2 (T (1:2)) is described. This T (1:2) and another sex-linked translocation involving chromosomes 1 and 3 (T (1:3)) have been identified by appropriate crosses to a marker strain, and by chromosome examinations of meiotic and mitotic tissues. The break-point in T (1:2) occurs at 0·3 of a cross-over unit from M on chromosome 1 and 1·6 cross-over units from the *spot-abdomen* locus on chromosome 2. That in T (1:3) is 0·4 of a unit from the *red-eye* (*re*) locus and 6·5–7·1 units from M on chromosome 1 and 0·6 of a unit from the *black tarsi* (*blt*) locus on chromosome 3 (McDonald and Rai, 1970a; Rai *et al.,* (1970). McDonald and Rai (1970b) then crossed the two translocation populations, and pro-duced an entity whose three chromosomes were composed of (a) parts of chromosomes 1 and 2, (b) parts of chromosomes 1 and 3, and (c) a "new" chromosome composed of parts of all three of the wild-type chromosomes.

While the simple translocations were responsible for something of the order of 50% sterility, the double heterozygote produced by crossing T (1:2) and T (1:3) gave 87·5% sterility (McDonald and Rai, 1971). These figures fed into a computerized mathematical model showed that a population with a reproductive potential of one would be eradicated in eleven generations, if the same number of T (1:2) translocation males were released on five successive occasions, that number being four times the number of wild males in the population at the start of the release. With T (1:3) males eradication would be achieved in six generations other things being equal, while with the double heterozygote elimination would result in only five generations with only four instead of five releases. However, when the reproductive potential was doubled, neither single translocation was capable of eradication. On the other hand six releases of a 4:1 ratio of double heterozygote males to wild-type males would be sufficient to achieve eradication in seven generations of a population capable of a 5-fold increase per generation but not of a population with an 8-fold rate of increase. As already pointed out, the use of such translocations in the heterozygous state would entail their prior separation from mixed stocks. While this can be done with the aid of suitable markers, it could hardly be done on a scale large enough to be a practical field proposition.

Out of more than 40 translocation stocks produced in *Ae. aegypti,* only two have so far been made homozygous (Lorimer *et al.,* 1972, and Rai and McDonald, 1972).These involve exchanges between chromosomes 1 and 3 (T (1:3)) and chromosomes 2 and 3 (T (2:3)). Surprisingly, both homozygotes were fully viable as shown by their 25% occurence in the offspring of heterozygote × heterozygote matings. However, their fertility was low (only 19% in T (1:3) and 55% in T (2:3) of eggs laid hatched). This has been attributed to the RED stock background incorporated during isolation, the RED stock being of reduced fertility (62% ± 21·0) compared with ROCK (84% ± 5·1) and wild populations (near 100%). It was thus realized that before these translocations could be used to control field populations, they would have to be crossed to wild-populations and re-isolated using a marker strain in which the marker genes were similarly transferred to a wild-type background. This would not only increase their fertility, but would also increase the likelihood of their being competitive in mating with the wild females. The stock, ROCK, from which they were derived has been "in colony" in the laboratory for more than 40 years. This is in fact now being done in Delhi[1]. In the meantime the original American stocks (now called T_1/T_1 and T_2/T_2) and their hybrids (the double heterozygotes T_1/T_2 and T_2/T_1) have been evaluated in cage competition trials in Delhi and a

[1] Information derived from monthly reports of the WHO/ICMR Research Unit on the Genetic Control of Mosquitoes, New Delhi, India.

field release of a simple male-linked translocation (T(1:2)) made to see if it would be incorporated into the wild population. That it was so was shown by the recovery of wild females exhibiting semi-sterility some weeks after the release (Rai et al., 1973). No attempt has yet been made to control a field population of Ae. aegypti by this method, however.

Comparisons between the two homozygous translocations, reciprocal double heterozygotes and the local Delhi stock show almost equal survival and speed of development in the aquatic stages, while the heterozygotes show a better male survival than the homozygotes or the Delhi strain. In cage mating competition experiments (T_1/T_1) did not perform very well, but the other three $(T_2/T_2, T_1/T_2$ and $T_2/T_1)$ were not significantly different from the Delhi strain. Male mating ability was slightly better in the heterozygotes than in the homozygotes or the wild strain. Cage releases of double heterozygote males at a proportion of 6:1:1 with Delhi males and females resulted in a 70% hatch of eggs laid by the females indicating poor competitiveness, and an 87% hatch in the F_2 generation showed that a rapid selective elimination of the translocation was occurring. A repeat of this experiment gave a better resultant egg hatch of 54%, and in fact 26 of 35 females gave reduced egg-hatches indicating they had been mated by doubly heterozygous males. However, the F_2 generation showed 73% hatch, again indicating rapid selective elimination of the translocations. It was later found that double heterozygote × wild-type matings did not always produce reduced fertility—in fact one-quarter of such matings proved fully fertile—indicating that free recombination between translocations was occurring.

C. *Culex pipiens*

Convinced that cytoplasmic incompatibility is a comparatively rare phenomenon and could only have a limited application in genetic control, Laven turned his attention to the isolation of translocations in *Culex pipiens* (Laven, 1969c, d; Laven and Jost, 1971; Laven et al., 1971a; Laven, 1972). Altogether 124 different translocation stocks have been isolated using semi-sterility as the sole criterion of recognition. Seven male-linked translocations have been selected to purity and one has been re-irradiated. From the second irradiation stable lines producing sterilities between 80 and 92% have been isolated. From 37 stocks containing female-linked and autosomal translocations 6 have been made homozygous and a male-linked translocation has been combined with two different homozygous ones to give a population 75% sterile on inbreeding, the males of which on outcrossing with normal females produce 98% sterility. The cytological appearance of some of these translocations has been described by Jost and Laven (1971).

As with the mechanism of cytoplasmic incompatibility Laven has been the first to use a translocation as a control measure in the field. He used a simple male-linked translocation in *C. p. pipiens* giving 50% sterility, and introduced it into an isolated population of the same species in a small village 16 km north of Montpellier in the south of France (Laven *et al.*, 1971b). The wild population was derived from a single breeding source (a well) and its output could be monitored with some degree of accuracy. Just before releases were made in the summer of 1970, some 2 400 mosquitoes were emerging each day but less than two weeks later this had increased to 21 000. In the meantime releases were made, but only reached a ratio of 1:1 one week after this peak output. From then on, however, an estimated 5:1 ratio of translocation: wild-type was maintained for six weeks. The first semi-sterile rafts appeared on the fifth day after releases started, and some three weeks later constituted more than 35% of rafts collected. Only three weeks after the release ratio had reached 5:1 the proportion of such rafts had reached 75%, and by the end of the two-month period of the experiment, which coincided with the commencement of hibernation, this figure was 95%. Returning in the middle of April of the following year Laven and his co-workers were still able to show that 89% of 420 egg-rafts laid by females coming out of hibernation were semi-sterile (Laven *et al.*, (1972), but that the total egg-rafts in sampling containers was very much lower than in the preceding year, and remained so until the end of July. Further observations by Cousserans and Guille (1972) showed this situation to continue until the end of the first week in September after which time and until the beginning of October the number of rafts was greater than in the preceding year. However, the average percentage of semi-sterile rafts was 50 even one year after the releases were made.

The integration of translocation and incompatibility in *C. pipiens* has already been referred to in the chapter on cytoplasmic incompatibility. The theory of the use of such a combined mechanism has been expounded by Laven and Aslamkhan (1970), and more recently by Laven (1972). The translocation system concerned is a male-linked multiple generating 85% sterility, in a cytoplasm bi-directionally incompatible with the population to be controlled. The release of both sexes of the integrated strain A into a wild population B will present the following mating and offspring possibilities:

A♂ × A♀	15% of the offspring hatch
A♂ × B♀ ⎫	incompatible and therefore
B♂ × A♀ ⎬	no offspring produced
B♂ × B♀	near 100% of the offspring hatch

If releases are made in numbers equal to those of the wild population, and if mating is at random, there will be a 50% reduction in B and a 92·5% reduction in A, and two more releases of similar numbers would theoretically result in the elimination of B and the replacement of it by A with its "built-in" 85% sterility. Such a population might be tolerated in terms of lower disease transmission and biting nuisance, though this is doubtful as density-dependence compensation could occur. If not, the release of a further integrated translocation strain incompatible with the first might eliminate the species altogether, especially if environmental conditions are becoming unfavourable. Laven (1972) sees a disadvantage in releasing females which can of course bite like the wild population, and transmit disease just as well possibly, but their effect would be short-lived. One big advantage is the lack of need of sex separation, and in fact Laven visualizes the creation of additional breeding places, and the seeding of them with the integrated strain.

With these ideas in mind, the D3/71 integrated strain of *C. pipiens* referred to in the chapter on cytoplasmic incompatibility has been released into a large walk-in cage in Delhi. 900 males and 900 females of this strain, and 100 males and 100 females of the local Delhi strain were added to the cage over a period of five consecutive days. Among 2 158 egg-rafts obtained, 3% gave normal hatches, 22% were sterile and 75% gave 5–25% hatch (the result of mating with translocated males). Expected figures assuming random mating were 1%, 18% and 81%, respectively. These figures indicate a lower competitiveness of the males of the release strain, and a tendency towards assortative (like-with-like) mating[22].

D3/71 has a European genome and might therefore be expected to be less competitive under Delhi conditions. Therefore, translocations isolated from the Delhi population itself have been combined with the D3 cytoplasm by crossing through intermediate strains from Bangkok and Fresno. A strain now exists (IS-31B), which consistently imparts 70% sterility, and is equally competitive with the wild Delhi population. This is being expanded with the intention of using it for field releases[1].

D. Other Culicine Mosquitoes

Among other culicine mosquitoes translocations have been isolated in *Ae. albopictus* (Laven *et al.*, (1971c), and *C. tritaeniorhynchus* (Laven *et al.*, (1971c) and Sakai *et al.*, 1971). The latter species, an important vector of the Japanese B encephalitis virus, has been particularly well studied from a genetics standpoint and though no homozygous translocations have yet been produced some 46 translocation stocks have been isolated (Sakai *et al.*, 1971).

[1] Information derived from monthly reports of the WHO/ICMR Research Unit for the Genetic Control of Mosquitoes, New Delhi, India.

E. Higher Diptera

Among the higher Diptera of medical and veterinary importance attempts have been made to isolate translocations in the tsetse-fly, *Glossina austeni,* the housefly, *Musca domestica,* the sheep blow-fly *Lucilia sericata,* and the screw-worm, *Cochliomyia hominivorax.* Curtis (1968a, 1970 and 1971a) descibes his methods with *G. austeni.* Without markers he was forced to rely on measurements of fertility. Some 300 individuals from seven different stocks were investigated, and in three of the stocks the presence of homozygotes indicated. However, there was evidence of reduced viability in these lines which may have been due to the translocations themselves or to the amount of inbreeding involved in producing them. There has now been cytological confirmation of some of these translocations (Curtis *et al.,* 1972). Curtis (1968a) is convinced that tsetse-flies would be good subjects for control with semi-sterility mechanisms, as they normally occur in low density in the wild, and show no major fluctuations in abundance indicating a relative absence of density-dependence compensation. In support of this is the evidence from attempted control of these insects by the use of insecticides dispersed from aircraft (Hocking *et al.,* 1953). Reductions of more than 90% of pre-spray numbers resulted, and recovery of population size was very slow to occur when control measures ceased.

Wagoner (1969) and Wagoner *et al.,* (1969 and 1971), give the results of their investigations of some 300 translocation lines produced in the housefly, *M. domestica.* Both autosomal-autosomal and Y-autosomal translocations have been isolated and three of the former made homozygous. The homozygous ones are simple translocations with low heterozygous sterility. Among autosomal heterozygotes, the average fertility of simple ones was 45%, of double ones 34% and triple ones 21%. Wagoner (1969) describes the release (into a large walk-in type cage) of a line heterozygous for multiple translocations involving chromosomes 2, 3 and 5 at a 9:1 ratio to the wild type. The wild population was reduced to 3·2% of the control level in one generation and to 0·25% in the second generation. Wagoner *et al.,* (1971) describe similar experiments using two strains A and B heterozygous for translocations between three autosomes and one strain C heterozygous for translocations between the Y chromosome and two of the autosomes. A and B were more efficient than C. At a 9:1 ratio, the former produced a fertility coefficient of from 10–22% of the level in controls in the first generation, and an accumulated coefficient of 1–6% through four generations. Equivalent figures for C were 42–46% and 34·5%.

Childress (1969) has isolated six translocation lines in *L. sericata* but no homozygotes have yet been recorded. The screw-worm, *C. hominivorax,* was apparently the first pest insect in which a translocation was described. This arose spontaneously, and the female heterozygote generated a higher degree of sterility than the male. The homozygote was apparently inviable (LaChance *et al.*, 1964).

F. Lepidoptera

The sterility in Lepidoptera after exposure to irradiation or to chemo-sterilants was at first attributed to the production of dominant lethals and to the direct effects on sperm as in other insects. Proverbs (1962a) was the first to note that the progeny of irradiated codling moths (*Laspeyresia pomonella*) were sterile. It is now known that Lepidoptera in general are highly resistant to irradiation from the point of view of the production of dominant lethals, because of the nature of their chromosomes; they have no localized centromeres. However, chromosome breakages readily occur, and chromosome pieces as well as translocations are able to pass through nuclear divisions without loss. In other words, a given sterilizing dosage produces fewer dominant lethals but more viable translocations in Lepidoptera than in other insects. To sterilize in the normal sense of the word requires very high doses of radiation—more than 20 000 r in the cabbage looper, *Trichoplusia ni,* according to North and Holt (1971)—and such high doses often lead to reductions in longevity and mating competitiveness. Thus a suggested method is to use sub-sterilizing doses, release the insects and rely on the sterility in the F_1 generation to control a wild population. As pointed out by Proverbs (1969), however, the damage caused by the released insects themselves might be unsustainable economically where high-priced crops are concerned.

North and Holt (1971) showed that the release of males of the cabbage looper, *T. ni,* irradiated at 15 000 r could produce more than 90% control under cage conditions, and was more effective than the release of fully sterile moths, even though these males were far from sterile. However, sex separation involves added costs to such a technique, and finding that inseminated females could be fully sterilized by exposures to the same irradiation dose, they proceeded to release both sexes (after allowing time for insemination and after irradiation) and obtained more than 90% control from a single release over two generations at a ratio of 9 irradiated insects to one wild. They envisage the setting up of large colony cages, allowing the moths to mate, waiting long enough to obtain the first batch of eggs, and then irradiating the remaining adult moths and releasing them. A continuous progression of cages would be kept and there would be no need for separate colony cages.

IV. Compound Chromosomes

A special type of translocation, a compound chromosome, is known in *Drosophila melanogaster*, and involves the exchange of whole chromosome arms between a homologous autosome pair. Both left arms attach to one centromere and both right arms to the other (see Fig. 16). Other insects, and especially mosquitoes with their few metacentric chromosomes, are considered good potential candidates for the isolation of such chromosomal exchanges (Foster *et al.*, 1972). These authors describe the method of producing such compound chromosomes in *D. melanogaster*. It involves the mating of irradiated females to males already heterozygous for whole-arm exchanges. What the likelihood of finding the latter is is not mentioned.

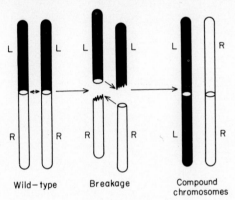

| Wild—type | Breakage | Compound chromosomes |

Fig. 16 Compound chromosome formation.

Compound chromosome strains of *D. melanogaster* breed true, and are fully viable, though their fertility is only 25–50% of the wild-type. When they mate with the wild-type, however, no viable offspring are produced. The hybrids die in the embryonic stage. Thus, theoretically, if the compound strain is released into a wild population in excess of four times the number of the latter, it should replace it within a small number of generations, and form an ideal transport mechanism for introducing genes advantageous to man. Actual cage competition experiments using a compound-second-chromosome strain and wild-type *D. melanogaster* have now confirmed these theoretical expectations (Childress, 1972).

One useful "spin-off" from all these investigations of translocations in insect pests has been the ability to identify for the first time actual chromosomes as the bearers of particular linkage-groups of mutant genes. This has now been accomplished in *Ae. aegypti* (McDonald and Rai, 1970a), *C. pipiens* (Jost and Laven, 1971), *C. tritaeniorhynchus* (Baker *et al.*, 1971), *M. domestica* (Wagoner, 1967) and in *L. cuprina* (Childress, 1969).

8. Other Methods of Genetic Control

I. Lethal Factors

Any deliberate alteration of the hereditary material of natural populations can be considered to come under the general heading of genetic control. The introduction of genes by straightforward overflooding without special carrier mechanisms is conceivable. These might be genes harmful to the insect population or genes altering some character of the insect in man's favour, such as a diversion of feeding habits. Natural populations habitually carry a considerable load of deleterious genes—straightforward lethal factors, conditional lethals and genes affecting viability. These are mostly recessive, but can be isolated by inbreeding in the laboratory. Released in the homozygous or even heterozygous condition, they would usually soon be eliminated by the normal processes of natural selection. Some carrier mechanism would be essential for their successful spread, and we have already considered the use of translocations as such. Another method now to be discussed involves the phenomenon of meiotic drive. Both are replacement mechanisms resulting in the substitution of one insect population by another. This might be carried even further to the point of replacing one harmful species by another harmless one. This will also be considered here.

The most likely candidates for introduction by themselves are dominant genes of the conditional lethal variety. These govern characters which have deleterious effects under certain conditions but not under others. Temperature-sensitive lethals come into this category. Insects carrying them can be mass-reared and released under conditions favourable to their survival in anticipation of a later seasonal change to a temperature lethal to them. They may of course be heat-sensitive or cold-sensitive. Other conditional lethals govern diapause, feeding and cocoon-formation. Non-diapausing insects could not survive unfavourable seasons. Non-feeding mutants might be mass-reared on artificial diets, but be unable to feed on natural food sources. Insects unable to construct functional cocoons, like those without the ability to diapause, would succumb through over-exposure to rigorous climatic conditions. Even the spread of these dominant conditional lethals will be assisted by their incorporation in some form of transport system.

II. Meiotic Drive and Sex Distortion

The phenomenon of meiotic drive has been recognized as a useful potential for the deliberate introduction of genes into populations. Sandler and Novitski (1957) first used this term to describe those occasions where unequal recoveries of alleles from heterozygotes occurred. An individual of constitution Aa producing more "A" gametes than "a" or vice versa is said to exhibit this phenomenon, and the explanation is that the one chromosome of a pair is "driven" more than the other at meiosis and is consequently more likely to end up in a functional gamete. *Segregation-Distorter (SD)* on chromosome 2 of *Drosophila melanogaster* is such an allele. Males heterozygous at this locus may pass the *SD*-bearing chromosome to more than 95% of their offspring (Sandler *et al.*, 1959). Novitski and Hanks (1961) later discovered another meiotic drive system distorting the sex-ratio in favour of females. This was called the *Recovery-Disrupter* system (*RD*), and male heterozygotes were here capable of producing progeny up to 67% female, because of a reduction in recovery of Y chromosomes (Hanks, 1964). A similar mechanism is indicated in the Lepidopteran *Acraea encedon* in equatorial Africa (Owen, 1970). Though this species is not an agricultural pest, the author recognizes the potentiality of sex-distortion as a means of genetic control. He found gross distortions in favour of females in nature in Sierra Leone, and showed there to be two sorts of females, one producing males and females, the other only females. Hamilton (1967) postulates the cause to be a driving Y chromosome (in Lepidoptera females are XY instead of the usual XX). It is suggested that in populations with normal sex ratios, the effects of this drive mechanism are covered up by suppressor genes. Sufficient and continuous sex-distortion could of course lead to population extinction, and Owen and Chanter (1969) describe such an occurrence in Uganda.

A. *Aedes Aegypti*

Sex-distortion in favour of males in an insect where the female is a disease transmitter would have the additional effect of reducing transmission before extinction is achieved. Such meiotic drive male distortion has in fact been found in *Aedes aegypti*. Craig *et al.*, (1960) were the first to notice that males from families already predominantly male produced high proportions of males in their progeny regardless of the female they were crossed to, and that this distortion was not due to any selective female mortality in the eggs, larvae or pupae (see also Hickey, 1970).

The mechanism of sex-determination in *Ae. aegypti* has already been referred to in the chapter on translocations. A single sex locus (m) on the smallest pair of chromosomes results in homozygotes (mm) being female and heterozygotes (Mm) being male. Hickey and Craig (1966a,b) postulate a *Distorter* factor (D) at or near this sex locus, which when in the heterozygous sex in the heterozygous genotype, $M^D m^d$, functions through a drive mechanism as a sex-distorter, producing more male-producing spermatozoa (M^D) than female (m^d). m^D is resistant to the effect of M^D. Laboratory population cage experiments involving different proportions of sex-distorting males ($M^D m^d$) and normal males ($M^D m^D$) put together with females ($m^d m^d$) and left to self-propagate for periods up to 43 weeks showed consistent sex-distortion as a rule, and in one case a marked population decline.

Sex-distorter factors are not uncommon in *Ae. aegypti* populations, and Hickey and Craig (1966a) suggest breeding schemes for their detection. There are also indications of differing degrees of distortion (as high as 96% males) dependent on a least four alleles at the d locus and two variants of D (Wood, personal communication)[1].

Wood has shown that natural populations are polymorphic at the d locus, ensuring that sex ratio distortion is not extreme and populations survive. But whether a low stable sex ratio will persist in the face of distorter factors introduced from other populations remains to be established.

There is also the possibility that balancing and stabilizing genes at other loci accumulate to offset extreme sex ratio distortion. In one laboratory experiment of Hickey and Craig (1966b) where each generation was separately reared, reversion to a normal sex ratio occurred in eleven generations. After fourteen generations distorter males could still be recovered in quantity, and such an accumulation of stabilizing genes was indicated. However, Wood[1] (personal communication) has pointed out that the same effect can arise from (a) selection of pre-existing variation at the d locus, (b) crossing over between M and d, (c) mutation of d to D and (d) contamination. Wood has carried out a similar experiment with no decline in sex ratio over ten generations.

Craig and Hickey (1967) have already suggested uses for such meiotic drive mechanism for genetic control purposes apart from its sex-distortion effect. A female-sterile mutant, *bronze*, is located near to the sex-locus (Bhalla and Craig, 1964), and could thus be introduced to hasten the elimination of natural populations. Other sex-linked genes are known, which affect the susceptibility of the mosquito to filarial infections (see Chapter 7). The effect of the introduction of the refractory alleles would be to lower the disease-transmission capability.

[1] Dr. R. J. Wood, Zoology Department, Manchester University.

B. *Musca Domestica*

A detailed study of the mechanisms of sex determination in the housefly (*M. domestica*) has revealed complicated combinations of XX and XY chromosomes, *M* and *m* autosomal alleles and the presence and partial or complete absence of what has been called an *F* factor (Milani, 1971). In the laboratory it has been possible to construct combinations of these factors, which when crossed give nothing but male offspring. These males are perfectly normal in their fertility, and may even show evidence of hybrid vigour. Thus it has been suggested that if the sex formula of a wild population is known, a certain combination constructed in the laboratory could be released as males which when they mated with the wild females would give only male offspring, thus distorting the sex ratio so much that population extinction might follow. Considered more reliable and practicable by Milani (1971), however, is the use of such mechanisms in the mass production of F_1 hybrid populations composed of males only, for sterilization and release—in other words, as a large-scale sex-separation technique. The meiotic drive sex-distortion mechanism just described in *Ae. aegypti* might be more reliably used in this manner too, though complete sex-distortion is rare in this case. Any sex distortion without accompanying differential sex mortality would be a help in the mass production of that one sex.

Another mechanism in the housefly which can result in the production of only male offspring has been discovered recently by McDonald (1971). This is a combination of an autosome inherited by male offspring only and a temperature-sensitive lethal factor. At 25·6°C such a strain produces both male and female progeny, but only males when the temperature is raised to 33·3°C.

III. Species Replacement

Perhaps not strictly deliberate genetic manipulation but nevertheless basically dependent on genetic differences is the idea of species competition for the same ecological niche leading to the possible replacement of a harmful pest by a harmless one. This might be assisted by the artificial release of mass-produced individuals of the desired species into areas where both species already exist or by the deliberate introduction of a foreign (allopatric) species. The latter is being considered as a means of replacing *Aedes polynesiensis*, the natural vector of aperiodic *Wuchereria bancrofti* in the South Pacific with the non-vector species *Ae. albopictus*. Gubler (1970a,b) showed that male *Ae. albopictus* readily mated with *Ae. polynesiensis* females in laboratory cages even in the presence of males of the latter species, and that such mated females laid sterile eggs. At a ratio of ten *Ae. albopictus* males to every *Ae. polynesiensis* male a cage colony of the latter species was eradicated. Ali and Rozeboom (1971a) extended these observations to a large room-size cage into which equal numbers of both sexes of each species were released. Mating pairs were later captured and identified. 72% of them were homologous *Ae. albopictus* matings. Only 5 of 222 *Ae. albopictus* females were found pairing with *Ae. polynesiensis* males, while 56 out of 78 *Ae. polynesiensis* females were found with *Ae. albopictus* males. Of the total males caught in these mating pairs 86% were in fact *Ae. albopictus*, a reflection of the greater mating activity of this species. When 400 virgin *Ae. polynesiensis* females, 400 virgin *Ae. albopictus* females and 400 *Ae. albopictus* males werd released together it was established by dissecting females five days after release that some. *Ae. polynesiensis* females (32 out of 100 examined) had in fact been inseminated. A sample of 50 *Ae. albopictus* females examined at the same time showed them all to have been inseminated.

Later the two species were allowed to compete freely in the large cage over a period of five months during which time the population size varied from 3 000–6 000 individuals. At the end of the period the proportion of *Ae. polynesiensis* had declined from an initial 40% down to about 5%. Periodic captures and identifications of mating pairs again showed most of the males to be *Ae. albopictus* even when caught with *Ae. polynesiensis* females. In spite of this, however, 85% of captured *Ae. polynesiensis* females laid fertile eggs as compared with 90% of *Ae. albopictus* females. Some partial hatches of eggs laid by *Ae. polynesiensis* may have been the result of double matings by males of both species. Thus cross-mating between the two species was demonstrated in a large cage allowing more freedom of escape if interspecific attentions were unwanted. However, in spite of the high frequency of this cross-mating, there was no apparent loss in fertility of *A. polynesiensis* females even when the proportion of this species was very low. An explanation for this may be that mating takes place in this species in a shorter time after emergence than in *Ae. albopictus*, and that *Ae. polynesiensis* females are inseminated by their own males before *Ae. albopictus* males become sexually active. Besides cross-mating, harassment of *Ae. polynesiensis* females by *Ae. albopictus* males leading to decreased feeding and longevity undoubtedly contributed to the decline of the former species.

Ali and Rozeboom (1971b) have also tested different populations of *Ae. albopictus* from Hawaii, Poona, Samoa, Tahiti and Taiaro in competition with different populations of *Ae. polynesiensis* from Samoa, Tahiti and Taiaro. Hawaiian *Ae. albopictus* males inseminated fewer *Ae. polynesiensis* females from all three areas than did those from Poona and the other places. Rozeboom (1971) then established a population of *Ae. polynesiensis* from Samoa in a large cage, and added male and female *Ae.*

albopictus from a Poona population to it. The former was soon replaced by the latter.

Aquatic stage competition experiments between the same two species were carried out by Lowrie (1973). *Ae. albopictus* larvae grew more rapidly and showed higher survivals than those of *Ae. polynesiensis*, and differences were more marked at higher densities of mixed species than at lower ones. Gubler (1972) showed that the two species offered different oviposition sites in the laboratory showed virtually no differences in preference.

9. Summary and Conclusion

It now seems evident that the use of non-specific insecticides, particularly in outdoor situations, presents an environmental hazard, indiscriminately killing harmful and beneficial organisms alike and leading to an accumulation of toxic residues. In any case insect resistance is a growing threat to their efficiency, and competitive economical substitute chemicals are becoming harder and harder to find. Existing methods of crop culture practices, the growing of resistant plants, environmental sanitation and biological control methods will undoubtedly have more important roles to play in the future but it is doubtful whether they can ever match the efficiency of some of the modern insecticides. New methods such as the use of specific attractants and juvenile hormones have not yet been adequately assessed.

Genetic control methods as well as being species specific and non-contaminating have distinct potentialities in their own right. In their simplest form they involve the controlled exposure of insects to those sublethal irradiation doses or chemosterilant concentrations which have the least effect on their subsequent activity, competitiveness and longevity but which are sufficient to induce dominant lethal mutations in most of the developing gametes. These insects are then released into wild populations and when they mate with wild individuals, union of gametes may follow, but because one of them bears a dominant lethal, death occurs at an early stage in zygotic development. Lower irradiation doses (or chemosterilant concentrations, though they are seldom used in practice) result in translocations of broken chromosome parts without, in some cases, an effect on viability but in the heterozygous state always an effect on fertility. The release of such individuals results in a self-propagating form of partial sterility which seems more likely to lead to population replacement than population elimination and could provide a means of transporting genes advantageous to man.

Besides these deliberate sterilizing procedures there are other methods which take advantage of natural incompatibilities between populations of the same species or which make use of sterile hybrid insects produced by crossing closely-related species. The first of these is only known in mosquitoes among pest insects and involves the release of one sex of an incompatible strain into the area occupied by another population of the same species. Hybrid sterility with its attendant hybrid vigour which may or may not include an ability to overcome species mating barriers is again restricted to a small number of insect pests. Finally, naturally occurring sex-distortion and meiotic drive mechanisms also have their potential uses in eradication or in population replacement.

The successful application of all these different methods of genetic control depends first of all on whether or not the pest to be controlled can be reared on a large scale and whether what comes out of this mass-rearing method will successfully compete with the wild population for mates.

It is inevitable that some changes in insect quality must occur as a result of mass-cultivation. One of the most likely changes is in mating behaviour because of the very fact that the self-perpetuating colony required as a base for large-scale rearing is usually housed in restricted conditions which may not allow normal mating behaviour. Other changes must arise from the fact that high yields are aimed at, at least much higher than occurs in the wild. Smaller, and probably different selective forces are involved allowing the survival of individuals which would normally die. It should be stressed that these changes will occur not just where an interspecific method like hybrid sterility is concerned but also in all the other intraspecific methods and many examples have been cited.

Sterilization procedures on top of mass-cultivation may further affect mating performance. Colonization and rearing changes might be overcome to some extent by regular additions of wild insects to the parent colony but whatever is done, differences must be expected. Perhaps the attitude of Gast (1968) is to be commended of almost assuming a poor performance of release material and compensating by mass over-releases of the order of hundreds of sterile insects to every normal one. Then other forces may come into play besides actual sterile matings—the flushing effect of Monro (1966)—the aggression factor of Baumhover (1965) and possibly occurring also with sterile hybrid males—and even a density dependent interference factor.

Whether to release one sex or both is firstly determined by the genetic method used. Cytoplasmic incompatibility as it is used for population elimination demands the release of one sex only as does the release of sex-linked translocations and sterile hybrids, unless both sexes are sterile. With other techniques the overriding considerations are the availability of a sexing method workable on a large scale and with a high degree of accuracy, whether one sex is polygamous and the other monogamous and whether one sex is more of a pest than the other. The ideal and most likely situation seems to be the release of polygamous males into populations where the female is monogamous. Autosomal translocations are best released as both sexes for the most rapid effect.

Genetic control methods demand for their efficient application a more detailed knowledge of insect ecology than most methods, particularly of the quantitive aspects. Essential is a knowledge of dispersal range in assessing the area of release, of female longevity in determining the minimum period of release and of population size in deciding the magnitude of release. When to release may be largely dependent on logistic considerations of when the required magnitude can be achieved allowing for certain levels of fitness and competitiveness. Most consider the ideal time to be when the population is at its lowest level and least likely to expand though it may be argued that at this time insects reared under conditions ideal for maximum yield of numbers cannot compete in the rigorous conditions responsible for the natural decline of the wild population. Perhaps the ideal is to integrate release with some other method of population reduction at the time of the year when release material is most likely to be effective. In this situation it would be of considerable advantage if the release insects were resistant to the other control method.

It is really too soon to make comparisons between the various methods of genetic control. The one really big success, the screw-worm campaign, may not be such a success after all, though economic savings must have been great over the past ten years and more. This used the technique of sterilization by irradiation which inexplicably seems to be going out of favour. One wonders whether this is just because of the scarcity and high initial cost of irradiation sources or because of the greater complexity of sterilizing procedure. Judging from the current literature chemosterilants are much more popular in spite of the dangers associated with their use and in spite of the fact that no really workable autosterilization procedure has been discovered.

Feasibility studies of genetic methods start in the laboratory and must eventually involve cage competition experiments between material intended for release and wild populations. These cage conditions must be such as to allow the normal mating behaviour of the wild insect to be really meaningful. No laboratory or field-cage study can take into account all the problems which may be encountered and sooner or later a pilot field trial becomes a necessity. Great care should be taken in the choosing of a site for such a trial as further progress in the use of the method may stand or fall on the outcome of a single trial which will invariably be much more expensive than the laboratory investigations. The site should be as closely representative of conditions normally met with as possible and not so small as to be little more than an extension of the field-cage type of experiment. Of vital importance is its isolation and freedom from immigration. Almost compulsorily it should be a literal island outside the flight range of the insect to be controlled and perhaps quarantined in addition. If this is not possible then an "island" situation can be simulated by treatments preventing breeding in peripheral barrier zones extended to the maximum dispersal range of the insect concerned. A detailed knowledge of seasonal fluctuation of the wild population must be available and releases planned to fill periods when they are expected to be competitive and which are long enough for effects to be shown.

Passing from small-scale pilot project to large-scale application is largely wandering into the realms of the unknown at this stage in the development of genetic control methods. We only really have the experience of the control of the screw-worm to go on and this involved an insect normally occurring in low numbers and apparently with a low capacity of increase. To many people the extension of such techniques to the control of insects with a known high rate of increase is inconceivable especially where such insects are spatially continuous over large areas. The solution may come from improvement in mass -rearingtechniques resulting in hitherto-undreamed-of numbers or from a concentration on those self-propagating techniques requiring comparatively few individuals.

References

Abasa, R. O. and Hansen, E. J. (1969). *J. econ. Ent.* **62,** 334–338.
Ailam, G. and Galun, R. (1967). *Ann. ent. Soc. Amer.* **60,** 41–43.
Akiyama, J. (1973). *Trans. roy. Soc. trop. Med. Hyg.* **67,** 440–441.
Ali, S. R. and Rozeboom, L. E. (1971a). *Mosquito News* **31,** 80–84.
Ali, S. R. and Rozeboom, L. E. (1971b). *J. med. Ent.* **8,** 263–265.
Altman, R. M. (1963). *Amer. J. Hyg.* **73,** 221–227.
Auerbach, C. and Robson, J. M. (1947a). *Proc. roy. Soc. Edin.* **62B,** 271–283.
Auerbach, C. and Robson, J. M. (1947b). *Proc. roy. Soc. Edin.* **62B,** 284–291.
Bailey, D. L., LaBrecque, G. C. and Whitfield, T. L. (1970). *J. econ. Ent.* **63,** 1451–1454.
Baker, R. H., French, W. L. and Kitzmiller, J. B. (1962). *Mosquito News* **22,** 16–17.
Baker, R. H., Sakai, R. K. and Mian, A. (1971). *Science* **171,** 585–587.
Baldwin, W. F. and Chant, G. D. (1971). *International Atomic Energy Agency Publication: STI/PUB/265* Vienna, pp. 469–474.
Barr, A. R. (1966). *Proc. 33 Ann. Conf. California Mosq. Control Ass.* 32–35.
Barr, A. R. (1969). *Proc. 37 Ann Conf. California Mosq. Control Ass.* 19–24.
Baumhover, A. H., Graham, A. J., Bitter, B. A., Hopkins, D. E., New, W. D., Dudley, F. H. and Bushland, R. C. (1955). *J. econ. Ent.* **48,** 462–466.
Baumhover, A. H. (1965). *J. econ. Ent.* **58,** 544–548.
Berryman, A. A. (1967). *Canad. Ent.* **99,** 848–865.
Bertram, D. S. (1963). *Trans. roy. Soc. trop. Med. Hyg.* **57,** 322–335.
Bertram, D. S. (1964). *Trans. roy. Soc. trop. Med. Hyg.* **58,** 296–317.
Bertram, D. S. and Gigliolo, M. E. C. (1963). *Trans. roy. Soc. trop. Med. Hyg.* **57,** 234.
Bertram, D. S., Srivastava, S. C. and Msangi, A. S. (1964). *J. trop. Med. Hyg.* **67,** 51–57.
Bhalla, S. G. and Craig, G. B. (1964). *Bull. ent. Soc. Amer.* **11,** 171.
Boesiger, E. (1972). *Indian Council med. Res. tech. Rep. Ser.* **20,** 66–76.
Bogyo, T. P., Berryman, A. A. and Sweeney, T. A. (1971). *International Atomic Energy Agency Publication: STI/PUB/281* Vienna, pp. 19–25.
Boller, E. (1972). *Entomophaga* **17,** 9–25.
Bransby-Williams, W. R. (1971). *E. Afr. med. J.* **48,** 68–75.
Brown, A. W. A. (1971). *In* "Pesticides in the Environment" (R. White-Stevens ed.) Vol. 1, pp. 475–552. Dekker, New York.

Brown, A. W. A. and Pal, R. (1971). *In* "Insecticide Resistance in Arthropods" (W. H. O. Monograph Series no. 38) Geneva.

Bryan, J. H. (1968). *Nature (Lond.)* **218,** 489.

Bryan, J. H. (1972). *Nature (Lond.)* **239,** 519–520.

Bryan, J. H. (1973a). *Trans. roy. Soc. trop. Med. Hyg.* **67,** 64–69.

Bryan, J. H. (1973b). *Trans. roy. Soc. trop. Med. Hyg.* **67,** 70–84.

Bryan, J. H. (1973c). *Trans. roy. Soc. trop. Med. Hyg.* **67,** 85–91.

Bryan, J. H. and Coluzzi, M. (1971). *Bull. Wld Hlth Org.* **45,** 266–267.

Burnham, C. R. (1962). Chapter IV *In* "Discussions in Cytogenetics". Burgess Press, Minneapolis.

Bushland, R. C. (1971). *International Atomic Energy Agency Publication: STI/PUB/265* Vienna, pp. 3–14.

Butt, B. A., Hathaway, D. O., White, L. D. and Howell, J. F. (1970). *J. econ. Ent.* **63,** 912–915.

Campion, D. G. (1972). *Bull. ent. Res.* **61,** 577–635.

Campion, D. G. and Lewis, C. T. (1971). *International Atomic Energy Agency Publication: STI/PUB/265* Vienna, pp. 183–202.

Carcavallo, R. U. and Carabajal, C. A. (1971). *International Atomic Energy Agency Publication: STI/PUB/265* Vienna, pp. 247–251.

Carson, R. (1962). *In* "Silent Spring". Houghton, Mifflin Co., Boston.

Cerf, D. C. and Georghiou, G. P. (1972). *Nature (Lond.)* **239,** 401–402.

Chambers, D. L., Spencer, N. R., Tanaka, N. and Cunningham, R. T. (1970). *International Atomic Energy Agency Publication: STI/PUB/276* Vienna, pp. 99–102.

Chang, S. C. and Bořkovec, A. B. (1964). *J. econ. Ent.* **57,** 488–490.

Childress, D. (1969). *Chromosoma (Berl.)* **26,** 208–214.

Childress, D. (1972). *Genetics* **72,** 183–186.

Christenson, L. D. (1966). U.S. Department of Agriculture, Agricultural Research Service Publication no. 33–110, pp. 95–102.

Coaker, T. H. and Smith, J. L. (1970). *Bull. ent. Res.* **60,** 53–59.

Coluzzi, M. (1968). *Parassitologia* **10,** 179–183.

Coluzzi, M. (1971). *Ann. Parasit.* **46,** 91–101.

Coluzzi, M. and Sabatini, A. (1967). *Parassitologia* **9,** 73–88.

Coluzzi, M. and Sabatini, A. (1968a). *Parassitologia* **10,** 155–165.

Coluzzi, M. and Sabatini, A. (1968b). *Riv. Parassit.* **29,** 49–70.

Coluzzi, M. and Sabatini, A. (1969). *Parassitologia* **11,** 177–187.

Coluzzi, M., Girioni, A. M. and Muir, D. A. (1970). *Parassitologia* **12,** 119–123.

Coluzzi, M., Cancrini, G. and DiDeco, M. (1971). *Parassitologia* **13,** 445–448.

Conway, G. R. (1970). *Misc. Publ. ent. Soc. Amer.* **7,** 181–191.

Corradetti, A., Dojmi di Delupis, G. L., Palmieri, C. and Piccione, G. (1970). *Parassitologia* **12,** 81–99.

Cousserans, J. and Guille, G. (1972). *Bull. Biol.* **106,** 337–343.

Craig, G. B. (1964). *Bull. Wld Hlth Org.* **31,** 469–473.

Craig, G. B. (1967). *Science* **156,** 1499–1501.

Craig, G. B. (1970). *Misc. Publ. ent. Soc. Amer* **7,** 130–133.

Craig, G. B. and Fuchs, M. S. (1969). *United States Patent Office Patent No. 3,450,816 of June 17, 1969.*

Craig, G. B. and Hickey, W. A. (1967). Chapter 3 *In* "Genetics of Insect Vectors of Disease" (J. W. Wright and R. Pal, eds.), pp. 67–131. Elsevier, Amsterdam.

Craig, G. B., Hickey, W. A. and VandeHey, R. C. (1960). *Science* **132**, 1887–1889.

Crystal, M. M. (1965). *J. med. Ent.* **2**, 317–319.

Cuellar, C. B. (1969a). *Bull. Wld Hlth Org.* **40**, 205–212.

Cuellar, C. B. (1969b). *Bull. Wld Hlth Org.* **40**, 213–220.

Cuellar, C. B. (1970). *Trans. roy. Soc. trop. Med. Hyg.* **64**, 475.

Cuellar, C. B. (1973a). *International Atomic Energy Agency Publication: STI/PUB/340* Vienna, pp. 5–15.

Cuellar, C. B. (1973b). *International Atomic Energy Agency Publication: STI/PUB/340* Vienna, pp. 149–163.

Cuellar, C. B. (1973c). *Parassitologia* **15**, 79–85.

Cuellar, C. B., Sawyer, B. and Davidson, G. (1970). *Trans. roy. Soc. trop. Med. Hyg.* **64**, 476.

Curtis, C. F. (1968a). *Bull. ent. Res.* **57**, 509–523.

Curtis, C. F. (1968b). *Nature (Lond.)* **218**, 368–369.

Curtis, C. F. (1968c). *J. insect Phys.* **14**, 1365–1380.

Curtis, C. F. (1970). *Proc. Ist. int. Symp. on Tsetse Fly Breeding, Lisbon.*

Curtis, C. F. (1971a). *International Atomic Energy Agency Publication: STI/PUB/265* Vienna, pp. 425–433.

Curtis, C. F. (1971b). *In* "Advances in Reproductive Physiology". Vol. 5, pp. 119–165. Academic Press, New York and London.

Curtis, C. F. (1972). *Acta. trop.* **29**, 250–268.

Curtis, C. F., and Hill W. G. (1971). *Theor. pop. Biol.* **2**, 71–90.

Curtis, C. F. and Robinson, A. S. (1971). *Genetics* **69**, 97–113.

Curtis, C. F., Southern, D. I., Pell, P. E. and Craig-Cameron, T. A. (1972). *Gen. Res.* **20**, 101–113.

Dame, D. A. (1968). *International Atomic Energy Agency Publication: STI/PUB/184* Vienna, pp. 23–24.

Dame, D. A. and Ford, H. R. (1968). *Bull. ent. Res.* **58**, 213–219.

Dame, D. A. and Schmidt, C. H. (1964). *J. econ. Ent.* **57**, 77–81.

Dame, D. A. and Schmidt, C. H. (1970). *Bull. ent. Soc. Amer.* **16**, 24–30.

Dame, D. A., Woodard, D. B., Ford, H. R. and Weidhaas, D. E. (1964). *Mosquito News* **24**, 6–14.

Davich, T. B. (1969). *International Atomic Energy Agency Publication: STI/PUB/224* Vienna, pp. 65–72.

Davidson, G. (1956). *Nature (Lond.)* **178**, 863–864.

Davidson, G. (1957). *Nature (Lond.)* **180**, 1333.

Davidson, G. (1963a). *Bull. Wld Hlth Org.* **28**, 25–33.

Davidson, G. (1963b). *Bull. Wld Hlth Org.* **29**, 177–184.

Davidson, G. (1964). *Riv. Malar.* **43**, 167–183.

Davidson, G. (1969a). *Bull. Wld Hlth Org.* **40**, 221–228.

Davidson, G. (1969b). *Cah. O.R.S.T.O.M. sér. Ent. méd. Parasitol.* **7**, 151–154.

Davidson, G. (1971). *Ann. Parasit. hum. comp.* **46**, 149–163.

Davidson, G. and Hunt, R. H. (1973). *Parassitologia* **15**, 121–128.

Davidson, G. and Jackson, C. E. (1961). *Bull. Wld Hlth Org.* **25**, 209–217.

Davidson, G., Paterson, H. E., Coluzzi, M., Mason, G. F., and Micks D. W. (1967). Chapter 6 *In* "Genetics of Insect Vectors of Disease" (J. W. Wright and R. Pal, eds.) pp. 211–250. Elsevier, Amsterdam.

Davidson, G., Odetoyinbo, J. A., Colussa, B. and Coz, J. (1970). *Bull. Wld Hlth Org.* **42,** 55–67.

Davies, M., Keiding, J. and von Hofstein, C. G. (1958). *Nature (Lond.)* **182,** 1816–1817.

Dean, G. J., Phelps, R. J. and Williamson, B. (1968). *International Atomic Energy Agency Publication: STI/PUB/184* Vienna, pp. 31–36.

Dennhofer, U. (1971). *Anz. Schädlingsk.* **44,** 84–91.

Dobrotworsky, N. U. (1955). *Proc. Linn. Soc. N.S.W.* **80,** 33–43.

Donnelly, J. (1965). *Proc. XII int. Congr. Ent.* (1964). 253–254,

Dyte, C. E. (1972). *Nature (Lond.)* **238,** 48–49.

Eyraud, M. and Mouchet, J. (1970). *Cah. O.R.S.T.O.M. sér. Ent. méd. Parasitol.* **8,** 69–82.

Fahmy, O. G. and Fahmy, M. J. (1964). *Trans. roy. Soc. trop. Med. Hyg.* **58,** 318–326.

Fay, R. W. and Morlan, H. B. (1959). *Mosquito News* **19,** 144–147.

Fossati, A., Stahl, J. and Granges, J. (1971). *International Atomic Energy Agency Publication: STI/PUB/281* Vienna, pp. 41–47.

Foster, G. G., Whitten, M. J., Prout, T. and Gill, R. (1972). *Science* **176,** 875–880.

Fuchs, M. S., Craig, G. B. and Hiss, E. A. (1968). *Life Sciences* **7,** 835–839.

Fye, R. L. and LaBrecque, G. C. (1971). *J. econ. Ent.* **64,** 973–974.

Fye, R. L., LaBrecque, G. C., Morgan, P. B. and Bowman, M. C. (1968). *J. econ. Ent.* **61,** 1578–1581.

Gardiner, B. O. C. and Maddrell, S. H. P. (1972). *Bull. ent. Res.* **61,** 505–515.

Gast, R. T. (1968). *International Atomic Energy Agency Publication: STI/PUB/185* Vienna, pp. 59–67.

Geier, P. W. (1969). *International Atomic Energy Agency Publication: STI/PUB/223* Vienna, pp. 33–41.

George, J. A. (1967). *Mosquito News* **27,** 82–86.

George, J. A. and Brown, A. W. A. (1967). *J. econ. Ent.* **60,** 974–978.

Georghiou, G. P. (1971). *In* "Agricultural Chemicals – Harmony or Discord for Food, People and the Environment" (J. E. Swift ed.) pp. 112–124. Univ. Calif. Div. Agr. Sci.

Georghiou, G. P. (1972). *Amer. J. trop. Med. Hyg.* **21,** 797–806

Gerberg, E. J., Gentry, J. W. and Diven, L. H. (1968). *Mosquito News* **28,** 342–346.

Gillies, M. T. (1961). *Bull. ent. Res.* **52,** 99–127.

Goma, L. K. H. (1963). *Nature (Lond.).* **197,** 99–100.

Gomez-Núñez, J. C. (1971). *International Atomic Energy Agency Publication: STI/PUB/265* Vienna, pp. 157–165.

Gouck, H. K., Meifert, D. W. and Gahan, J. B. (1963). *J. econ. Ent.* **56,** 445–446.

Grant, G. G., Carmichael, A. G., Smith, C. N. and Brown, A. W. A. (1970). *J. econ. Ent.* **63,** 648– 650.

Green, C. A. (1972). *Ann. trop. Med. Parasit.* **66,** 143–147.

Grosch, D. S. and Sullivan, R. L. (1954). *Radiat. Res.* **1,** 294–320.

Grover, K. K. and Pillai, M. K. K. (1970a). *J. med. Ent.* **7**, 198–204.

Grover, K. K. and Pillai, M. K. K. (1970b). *Bull. Wld Hlth Org.* **42**, 807–815.

Gubler, D. J. (1970a). *J. med. Ent.* **7**, 65–70

Gubler, D. J. (1970b). *J. med. Ent.* **7**, 229–235.

Gubler, D. J. (1972). *J. med. Ent.* **8**, 675–682.

Gwadz, R. W., Craig, G. B. and Hickey, W. A. (1971). *Biol. Bull.* **140**, 201–214.

Hackett, L. W. (1937). *In* "Malaria in Europe". Oxford University Press, London.

Hamilton, W. D. (1967). *Science* **156**, 477–488.

Hanks, G. D. (1964). *Genetics* **50**, 123–130.

Hansens, E. J. (1965). *J. econ. Ent.* **58**, 944–946.

Hansens, E. J. and Granett, P. (1965).. *J econ. Ent.* **58**, 157–158.

Hazard, E. I., Lofgren, C. S., Woodard, D. B., Ford, H. R. and Glancey, B. M. (1964). *Science* **145**, 500–501.

Henneberry, T. J. (1971). *International Atomic Energy Agency Publication: STI/PUB/281* Vienna, pp. 51–63.

Henneberry, T. J., Howland, A. F. and Wolf, W. W. (1965). *Proc. Conf. Electromagnetic Radiation Agric.* 34.

Hickey, W. A. (1970). *J. med. Ent.* **7**, 727–735.

Hickey, W. A. and Craig, G. B. (1966a). *Canad. J. Genet. Cytol.* **8**, 260–278.

Hickey, W. A. and Craig, G. B. (1966b). *Genetics* **53**, 1177–1196.

Hightower, B. G. and Graham, O. H. (1968). *International Atomic Energy Agency Publication: STI/PUB/184* Vienna, pp. 51–54.

Hocking, K. S., Parr, H. C. M., Yeo, D. and Anstey, D. (1953). *Bull. ent. Res.* **44**, 627–631.

Hogan, T. W. (1971). *Bull. ent. Res.* **60**, 383–390.

Hooper, G. H. S. (1970). *International Atomic Energy Agency Publication: STI/PUB/276* Vienna, pp. 3–12.

Hooper, G. H. S. (1971). *International Atomic Energy Agency Publication: STI/PUB/265* Vienna, pp. 87–95.

Hooper, G. H. S. and Nadel, D. J. (1970). *International Atomic Energy Agency Publication: STI/PUB/276* Vienna, pp. 51–58.

Horber, E. (1963). *International Atomic Energy Agency Publication: STI/PUB/74* Vienna, pp. 313–331.

Howland A. F. Vail, P. and Henneberry, T. J. (1966). *J. econ. Ent.* **59**, 194–196.

Huff, C. G. (1931). *J. prev. Med. (Baltimore)* **5**, 249–259.

Hunt, R. H. and Krafsur, E. S. (1972). *Trans. roy. Soc. trop. Med. Hyg.* **66**, 23–24.

Husseiny, M. M. and Madsen, H. F. (1964). *Helgardia* **36**, 113–137.

Itard, J. (1971). *Ann. Parasit. hum. comp.* **46**, 35–66.

Jermy, T. and Nagy, B. (1969). *International Atomic Energy Agency Publication: STI/PUB/224* Vienna, pp. 91–95.

Jones, J. C. (1973). *Nature (Lond.)* **242**, 343–344.

Jones, M. D. R. (1964). *J. insect Physiol.* **10**, 343–351.

Jordan, A. M. and Curtis, C. F. (1968). *Bull. ent. Res.* **58**, 399–410.

Jordan, A. M. and Curtis, C. F. (1972). *Bull. Wld Hlth Org.* **46**, 33–38.

Jost, E. (1970a). *Theo. Appl. Gen.* **40**, 251–256.

Jost, E. (1970b). *Wilhelm Roux' Arch. Entwickl. Mech. Org.* **166**, 173–188.

Jost, E. and Laven, H. (1971). *Chromosoma (Berl.)* **35**, 184–205.

Judson, C. L. (1967). *Ent. exp. appl.* **10**, 387–394.

Keiser, I., Steiner, L. F. and Kamasaki, H. (1965). *J. econ. Ent.* **58**, 682–685.

Kilama, W. L. and Craig, G. B. (1969). *Ann. trop. Med. Parasit.* **63**, 419–432.

Kitzmiller, J. B., Frizzi, G. and Baker, R. H. (1967). Chapter 5 *In* "Genetics of Insect Vectors of Disease" (J. W. Wright and R. Pal eds.) pp. 151–210. Elsevier, Amsterdam.

Klassen, W. and Matsamura, F. (1966). *Nature (Lond.)* **209**, 1155–1156.

Knipling, E. F. (1967). Chapter 20 *In* "Genetics of Insect Vectors of Disease" (J. W. Wright and R. Pal eds.) pp. 587–616. Elsevier, Amsterdam.

Knipling, E. F. (1972). *Proc. 14th. int. Congr. Ent.* Canberra, 4–18.

Kojima, K. (1971). *International Atomic Energy Agency Publication: STI/PUB/265* Vienna, pp. 477–487.

Krafsur, E. S. (1972). *Trans. roy. Soc. trop. Med. Hyg.* **66**, 22–23.

Krishnamurthy, B. S. (1972). *Indian Council med. Res. tech. Rep. Ser.* **20**, 138–143.

Krishnamurthy, B. S., Ray, S. N. and Joshi, G. C. (1962). *Indian J. Malar.* **16**, 365–373.

LaBrecque, G. C. (1972). *Indian Council med. Res. tech. Rep. Ser.* **20**, 123–125.

LaBrecque, G. C. and Keller, J. C. (eds.) (1965). *International Atomic Energy Agency Publication: Tech. Rep. Ser.* **44**, 1–79.

LaBrecque, G. C. and Meifert, D. W. (1966). *J. med. Ent.* **3**, 323–326.

LaBrecque, G. C. and Smith, C. N. (eds.) (1968). *In* "Principles of Insect Chemosterilization". North-Holland Publ. Co., Amsterdam.

LaBrecque, G. C., Meifert, D. W. and Smith, C. N. (1962a). *Science* **136**, 388–389.

LaBrecque, G. C., Smith, C. N. and Meifert, D. W. (1962b). *J. econ. Ent.* **55**, 449–451.

LaBrecque, G. C., Meifert, D. W. and Fye, R. L. (1963). *J. econ. Ent.* **56**, 150–152.

LaBreque, G. C., Meifert, D. W. and Rye, J. (1972a). *Canad. Ent.* **104**, 885–887.

LaBrecque, G. C., Bowman, M. C., Patterson, R. S. and Seawright, J. A. (1972b). *Bull. Wld Hlth Org.* **47**, 675–676.

LaChance, L. E. (1962). *Proc. N. cent. Br. ent. Soc. Amer.* **17**, 4pp.

LaChance, L. E. (1963). *Int. J. Radiat. Biol.* **7**, 321–331.

LaChance, L. E. (1967). Chapter 21 *In* "Genetics of Insect Vectors of Disease" (J. W. Wright and R. Pal eds.) pp. 617–650. Elsevier, Amsterdam.

LaChance, L. E. and Riemann, J. G. (1964). *Mutation Research* **1**, 318–333.

LaChance, L. E., Reimann, J. G. and Hopkins, D. E. (1964). *Genetics* **49**, 959–972.

Laird, M. (1967). *Nature (Lond.)* **216**, 1358.

Langley, P. A. and Abasa, R. O. (1970). *Ent. exp. appl.* **13**, 141–152.

Laven, H. (1951). *Evolution* **5**, 370–375.

Laven, H. (1957). *Z. Vererbungsl.* **88**, 478–516.

Laven, H. (1959). *Cold Spr. Harb. Symp. quant. Biol.* **24**, 166–173.

Laven, H. (1967a) Chapter 7 *In* "Genetics of Insect Vectors of Disease" (J. W. Wright and R. Pal eds.) pp. 251–275. Elsevier, Amsterdam.

Laven, H. (1967b). *Nature (Lond.)* **216**, 383–384.

Laven, H. (1969a). *Mosquito News* **29**, 70–74.

Laven, H. (1969b). *Mosquito News* **29**, 74–83.

Laven, H. (1969c). *Nature (Lond.)* **221**, 958–959.

Laven, H. (1969d). *Anz. Schädlingsk.* **42**, 17–19.

Laven, H. (1971). *Ann. Parasit.* **46**, 147–148.

Laven, H. (1972). *Indian Council med. Res. tech. Rep. Ser.* **20**, 95–109.

Laven, H. and Aslamkhan, M. (1970). *Pak. J. Sci.* **22**, 303–312.

Laven, H. and Jost, E. (1971). *Experientia (Basel)* **27**, 471–473.

Laven, H., Meyer, E., Bienok, R., Guille, G. and Ohmann, J. (1971a). *Experientia (Basel)* **27**, 968–969.

Laven, H., Cousserans, J. and Guille, G. (1971b). *Experientia (Basel)* **27**, 1355–1357.

Laven, H., Jost, E., Meyer, H. and Selinger, R. (1971c). *International Atomic Energy Agency Publication: STI/PUB/265* Vienna, pp. 415–424.

Laven, H., Coussearns, J. and Guille, G. (1972). *Nature (Lond.)* **236**, 456–457.

Leahy, M. G. and Craig, G. B. (1965). *Mosquito News* **25**, 448–452.

Lindquist, A. W. (ed.) (1963). *International Atomic Energy Agency Publication: Tech. Rep. Ser. No. 21,* Vienna, pp. 1–59.

Lorimer, N., Halliman, E. and Rai, K. S. (1972). *J. Hered.* **63**, 159–166.

Lowrie, R. C. (1973). *J. med. Ent.* **10**, 23–30.

Luckmann, W. H., Gangrade, G. and Broersma, D. B. (1967). *J. econ. Ent.* **60**, 737–741.

Macdonald, W. W. (1967). Chapter 19 *In* 'Genetics of Insect Vectors of Disease" (J. W. Wright and R. Pal, eds.) pp. 567–584. Elsevier, Amsterdam.

Macdonald, W. W., Sebastian, A. and Tun, M. M. (1968). *Ann. trop. Med. Parasit.* **62**, 200–209.

Mackauer, M. (1972). *Entomophaga* **17**, 27–48.

MacLeod, J. and Donnelly, J. (1961). *Ent. exp. appl.* **4**, 101–118.

Magaudda, P. L., Sacca, G. and Guarniera, D. (1969). *Ann. Inst. Sup. San.* **5**, 29–38.

Maksimović, M. (1971). *International Atomic Energy Agency Publication: STI/PUB/281* Vienna, pp. 75–80.

Manning, A. (1961). *Animal Behaviour* **9**, 82–92.

Martinez-Palacios, A. and Davidson, G. (1967). *Mosquito News* **27**, 55–56.

Masner, P., Slama, K. and Land, V. (1968). *Nature (Lond.)* **219**, 395–396.

Matsuzawa, H. and Fuji'i, Y. (1968). *Jap. J. San. Zool.* **19**, 210–212.

McClelland, G. A. H. (1967). Chapter 8 *In* "Genetics of Insect Vectors of Disease" (J. W. Wright and R. Pal, eds.) pp. 277–311. Elsevier, Amsterdam.

McDonald, I. C. (1971). *Science* **172**, 489.

McDonald, P. T. and Rai, K. S. (1970a). *Genetics* **66**, 475–485.

McDonald, P. T. and Rai, K. S. (1970b). *Science* **168**, 1229–1230.

McDonald, P. T. and Rai, K. S. (1971). *Bull. Wld Hlth Org.* **44**, 829–845.

Meifert, D. W. and LaBrecque, G. C. (1971). *J. med. Ent.* **8**, 43–45.

Meifert, D. W., LaBrecque, G. C., Smith, C. N. and Morgan, P. B. (1967a). *J. econ. Ent.* **60**, 480–485.

Meifert, D. W., Morgan, P. B. and LaBrecque ,G. C. (1967b). *J. econ. Ent.* **60**, 1336–1338.

Mellado, L. (1971). *International Atomic Energy Agency Publication: STI/PUB/265* Vienna, pp. 49–53.

Milani, R. (1971). *International Atomic Energy Agency Publication: STI/PUB/265* Vienna, pp. 381–396.

Monro, J. (1966). *Science,* **151,** 1536–1538.

Morgan, P. B. (1967). *J. econ. Ent.* **60,** 612–613.

Morgan, P. B., LaBrecque, G. C., Smith, C. N., Meifert, D. W. and Murvosh, C. M. (1967). *J. econ. Ent.* **60,** 1064–1067.

Morlan, H. B., McCray, E. M. and Kilpatrick, J. W. (1962). *Mosquito News* **22,** 295–300.

Mulla, M. S. (1964). *Mosquito News* **24,** 212–217.

Murray, W. S. and Bickley, W. E. (1964). *Maryland agric. exp. Sta. Bull.* **A–134,** 1–37.

de Murtas, I. D., Cirio, U., Guerrieri, G. and Enkerlin-S, D. (1970). *International Atomic Energy Agency Publication: STI/PUB/276* Vienna, pp. 59–70.

Nadel, D. J. and Peleg, B. A. (1968). *International Atomic Energy Agency Publication: STI/PUB/185* Vienna, pp. 87–90.

Nature (1973). News and Views. *Nature (Lond.)* **242.** 493–494.

Nash, T. A. M., Jordan, A. M. and Trewern, M. A. (1971). *International Atomic Energy Agency Publication: STI/PUB/265* Vienna, pp. 99–108.

Noordink, J. Ph. W. (1971). *International Atomic Energy Agency Publication: STI/PUB/265* Vienna, pp. 323–328.

North, D. T. and Holt, G. G. (1971). *International Atomic Energy Agency Publication: STI/PUB/281* Vienna, pp. 99–111.

Novitski, E. and Hanks, G. D. (1961). *Nature (Lond.)* **190,** 989–990.

Orphanidis, P. S. and Kalmoukos, P. E. (1971). *International Atomic Energy Agency Publication: STI/PUB/265* Vienna, pp. 55–65.

Orphanidis, P. S., Patsacos, P. G. and Kalmoukos, P. E. (1966). *Ann. Inst. phytopath. Benaki N. S.* **7,** 177–190.

Ouye, M. T., Graham, H. M., Garcia, R. S. and Martin, D. F. (1965). *J. econ. Ent.* **58,** 927–929.

Owen, D. F. (1970). *Nature (Lond.)* **225,** 662–663.

Owen, D. F. and Chanter, D. O. (1969). *J. Zool.* **157,** 345–374.

Paterson, H. E. (1964). *Riv. Malar.* **43,** 191–196.

Patterson, J. W. (1971). *Nature (New Biology, Lond.)* **233,** 176–177.

Patterson, R. S. and Lofgren, C. S. (1968). *Proc. 55th. ann. Meeting N. J. Mosq. Exterm. Ass.* 170–175.

Patterson, R. S., Lofgren, C. S. and Boston, M. D. (1967). *J. econ. Ent.* **60,** 1673–1675.

Patterson, R. S., Lofgren, C. S. and Boston, M. D. (1968a). *Mosquito News* **28,** 540–544.

Patterson, R. S., Lofgren, C. S. and Boston, M. D. (1968b). *The Florida Entomologist* **51,** 77–82.

Patterson, R. S., Ford, H. R., Lofgren, C. S. and Weidhaas, D. E. (1970a). *Mosquito News* **30,** 23–27.

Patterson, R. S., Weidhaas, D. E., Ford, H. R. and Lofgren, C. S. (1970b). *Science* **168,** 1368–1370.

Patterson, R. S., Boston, M. D., Ford, H. R. and Lofgren, C. S. (1971). *Mosquito News* **31,** 85–90.

Patterson, R. S., Boston, M. D. and Lofgren, C. S. (1972a). *Mosquito News* **32,** 95–98.

Patterson, R. S., Boston, M. D. and Lofgren, C. S. (1972b). *Mosquito News* **32,** 230–233.

Pausch, R. D. (1971). *J. econ. Ent.* **64,** 1462–1465.

Pausch, R. D. (1972). *J. econ. Ent.* **65,** 449–450.

Perlowagora-Szumlewicz, A. and Correia, M. V. (1972). *Rev. Inst. Med. trop. S. Paulo* **14,** 360–371.

Perry, J. (1950). *Ann. ent. Soc. Amer.* **43,** 123–136.

Pillai, M. K. K. and Grover, K. K. (1969) *Bull. Wld Hlth Org.* **40,** 229–233.

Plapp, F. W., Bigley, W. S., Chapman, G. A. and Eddy, G. W. (1962). *J. econ. Ent.* **55,** 607–613.

Potts, W. H. (1958). *Ann. trop. Med. Parasit.* **52,** 484–515.

Powell, J. R. and Craig, G. B. (1970). *J. econ. Ent.* **63,** 745–748.

Proverbs, M. D. (1962a). *Proc. ent. Soc. Ont.* **92,** 5–11.

Proverbs, M. D. (1962b). *Nature (Lond.)* **194,** 1297–1298.

Proverbs, M. D. (1969). *Ann. Rev. Ent.* **14,** 81–102.

Proverbs, M. D. (1971). *International Atomic Energy Agency Publication: STI/ PUB/281* Vienna, pp. 117–133.

Proverbs, M. D. and Newton, J. R. (1962). *J. econ. Ent.* **55,** 934–936.

Rabbani, M. G. and Kitzmiller, J. B. (1972). *Mosquito News* **32,** 421–432.

Rai, K. S. and Asman, M. (1968). *Proc. XII int. Congr. Genetics* **1,** 164.

Rai, K. S. and McDonald, P. T. (1972). *Indian Council med. Res. tech. Rep. Ser.* **20,** 77–94.

Rai, K. S., McDonald, P. T. and Asman, M. (1970). *Genetics* **66,** 635–651.

Rai, K. S., Grover, K. K. and Suguna, S. G. (1973). *Bull. Wld Hlth Org.* **48,** 49–56.

Rajagopalan, P. K., Yasuno, M. and LaBrecque, G. C. (1973). *Bull. Wld Hlth Org.* **48,** 631–635.

Ramakrishna, S. P., Krishnamurthy, B. S. and Singh, N. N. (1963). *Indian J. Malar.* **17,** 119–121.

Ramsdale, C. D. and Leport, G. H. (1967). *Bull. Wld Hlth Org.* **36,** 494–500.

Rapoport, I. A. (1947). *Dokl. vses. Akad. sel'.-khoz. Nauk.* **12,** 12–15 (*Chem Abstr.* **42,** 3790).

Rhode, R. H. (1970). *International Atomic Energy Agency Publication: STI/PUB/276* Vienna, pp. 43–50.

Riemann, J. G. and Thorson, B. J. (1969). *Ann. ent. Soc. Amer.* **62,** 828–834.

Rivosecchi, J. (1962). *Riv. Parassit.* **23,** 71–74.

Rozeboom, L. E. (1971) *Amer. J. trop. Med. Hyg.* **20,** 356–362.

Rozeboom, L. E. and Gilford, B. N. (1954). *Amer. J. Hyg.* **60,** 117–134.

Sacca, G. and Stella, E. (1964). *Riv. Parassit.* **25,** 279–294.

Sacca, G., Magrone, R. and Scirocchi, A. (1965). *Riv. Parassit.* **26,** 61–66.

Sacca, G., Scirocchi, A., De Meo, G. M. and Mastrilli, M. L. (1966a). *Atti. Soc. pelorit. Sci. fis. mat. nat.* **12,** 457–464.

Sacca, G., Scirocchi, A., Stella, E., Mastrilli, M. L. and De Meo, G. M. (1966b). *Atti. Soc. pelorit. Sci. fis. mat. nat.* **12,** 447–456.

Sacca, G., Stella, E., Mastrilli, M. L. and Gandoleo, D. (1969). *Parassitologia* **11**, 271–275.

Sakai, R. K., Baker, R. H. and Mian, A. (1971). *J. Hered,* **62**, 90–100.

Sandler, L. and Novitski, E. (1957). *Amer. Naturalist* **91**, 105–110.

Sandler, L., Hiraizumi, Y. and Sandler, I. (1959). *Genetics* **44**, 233–250.

Schmidt, C. H., Dame, D. A. and Weidhaas, D. E. (1964). *J. econ. Ent.* **57**, 753–756.

Serebrovsky, A. S. (1940). *Zool. Zh.* **19**, 618–630.

Serebrovsky, A. S. (1969). *International Atomic Energy Agency Publication: STI/PUB/224* Vienna, pp. 123–137.

Sharma, V. P., Patterson, R. S. and Ford, H. R. (1972). *Bull. Wld Hlth Org.* **47**, 429–432.

Sharma, V. P., Patterson, R. S., Grover, K. K. and LaBrecque, G. C. (1973). *Bull. Wld Hlth Org.* **48**, 45–48.

Shaw, J. G. and Sanchez Riviello, M. (1965). *J. econ. Ent.* **58**, 26–28.

Smith, C. N. (1966) "Insect Colonisation and Mass Rearing". Academic Press, New York and London.

Smith, C. N. (1967). Chapter 22 *In* "Genetics of Insect Vectors of Disease" (J. W. Wright and R. Pal, eds.) pp. 653–672. Elsevier, Amsterdam.

Smith, C. N., LaBrecque, G. C. and Bořkovec, A. B. (1964). *Ann. Rev. Ent.* **9**, 269–284.

Smith, R. H. (1971). *International Atomic Energy Agency Publication: STI/PUB/265* Vienna, pp. 453–465.

Smith, R. H. and von Borstel, R. C. (1972). *Science* **178**, 1164–1174.

Smith-White, S. and Woodhill, A. R. (1955). *Proc. Linn. Soc. N.S.W.* **79**, 163–176.

Snow, J. W., Burton, R. L., Sparks, A. N. and Cantelo, W. W. (1971). *International Atomic Energy Agency Publication: STI/PUB/285* Vienna, pp. 141–150.

Soerono, M., Davidson, G. and Muir, D. A. (1965). *Bull. Wld Hlth Org.* **32**, 161–168.

Stahler, N. (1971). *Ann. ent. Soc. Amer.* **64**, 1247–1249.

Stahler, N. and Terzian, L. A. (1966). *Ann. ent. Soc. Amer.* **59**, 763–765.

Stahler, N. and Terzian, L. A. (1968). *J. econ. Ent.* **61**, 1447.

Steiner, L. F., Mitchell, W. C. and Baumhover, A. H. (1962). *Int. J. appl. Radiat.* **13**, 427–434.

Steiner, L. F., Harris, E. J., Mitchell, W. C., Fujimoto, M. S. and Christenson, L. D. (1965a). *J. econ. Ent.* **58**, 519–522.

Steiner, L. F., Mitchell, W. C., Harris, E. J., Kozuma, T. T. and Fujimoto, M. S. (1965b). *J. econ. Ent.* **58**, 961–964.

Steiner, L. F., Hart, W. C., Harris, E. J., Cunningham, R. T., Ohinata, K. and Kamakahi, D. C. (1970). *J. econ. Ent.* **63**, 131–135.

Sudia, W. D. and Chamberlain, R. W. (1962). *Mosquito News* **22**, 126–128.

Terzian, L. A. and Stahler, N. (1966). *Radiat. Res.* **28**, 643–646.

Ticheler, J. (1971). *International Atomic Energy Agency Publication: STI/PUB/265* Vienna, pp. 59–67.

VandeHey, R. C. and Craig, G. B. (1958). *Bull. ent. Soc. Amer.* **4**, 102.

Vanderplank, F. L. (1947). *Trans. roy. ent. Soc. Lond.* **98**, 1–18.

Vanderplank, F. L. (1948). *Ann. trop. Med. Parasit.* **42,** 131–152.

Wagoner, D. E. (1967). *Genetics* **57,** 729–739.

Wagoner, D. E. (1969). *Bull. ent. Soc. Amer.* **15,** 220.

Wagoner, D. E., Nickel, C. A. and Johnson, O. A. (1969). *J. Hered.* **60,** 301–304.

Wagoner, D. E., Johnson, O. A. and Nickel, C. A. (1971). *Nature (Lond.)* **234,** 473–475.

Walliker, D., Carter, R. and Morgan, S. (1971). *Nature (Lond.)* **232,** 561–562.

Ward, R. A. (1963). *Exp. Parasit.* **13,** 328–341.

Weathersby, A. B. (1963). *Mosquito News* **28,** 249–251.

Weidhaas, D. E. (1968). Chapter 6 *In* "Principles of Insect Chemosterilization" (G. C. LaBrecque and C. N. Smith, eds.) pp. 275–314. Appleton-Century-Crofts, New York.

Weidhaas, D. E. and LaBrecque, G. C. (1970). *Bull. Wld Hlth Org.* **43,** 721–725.

Weidhaas, D. E. and Schmidt, C. H. (1963). *Mosquito News* **23,** 32–34.

Weidhaas, D. E., Schmidt, C. H. and Seabrook E. L. (1962). *Mosquito News* **22,** 283–291.

Weidhaas, D. E., Patterson, R. S., Lofgren, C. S. and Ford, H. R. (1971). *Mosquito News* **31,** 177–182.

Weidhaas, D. E., LaBrecque, G. C., Lofgren, C. S. and Schmidt, C. H. (1972). *Bull. Wld Hlth Org.* **47,** 309–315.

White, G. B. (1966). *Nature (Lond.)* **210,** 1372–1373.

White, G. B. (1970). *J. med. Ent.* **7,** 374–375.

White, G. B., Magayuka, S. A. and Boreham, P. F. L. (1972). *Bull. ent. Res.* **62,** 295–317.

Whitten, M. J. (1969). *J. econ. Ent.* **62,** 272–273.

Whitten, M. J. (1970). *In* "Concepts of Pest Management" (R. L. Rabb and F. E. Guthrie, eds.) pp. 119–135. N. C. State University, Raleigh.

Whitten, M. J. (1971a). *Science* **171,** 682–684.

Whitten, M. J. (1971b). *International Atomic Energy Agency Publication: STI/PUB/265* Vienna, pp. 399–410.

Whitten, M. J. (1971c). *J. econ. Ent.* **64,** 1310–1311.

Whitten, M. J. and Norris, K. R. (1967). *Nature (Lond.)* **216,** 1136.

Whitten, M. J. and Taylor, W. C. (1970). *J. econ. Ent.* **63,** 269–272.

Woodhill, A. R. (1949a). *Proc. Linn. Soc. N.S.W.* **74,** 19–20.

Woodhill, A. R. (1949b). *Proc. Linn. Soc. N.S.W.* **74,** 140–144.

Woodhill, A. R. (1950). *Proc. Linn. Soc. N.S.W.* **75,** 251–253.

Woodhill, A. R. (1954). *Proc. Linn. Soc. N.S.W.* **79,** 19–20.

World Health Organization (1964). *Techn. Rep. Ser.* **268.**

World Health Organization (1971). *Chron. Wld Hlth Org.* **25,** 230–235.

Yasuno, M. (1972). *Indian Council med. Res. tech. Rep. Ser.* **20,** 144–148.

Yen, J., H. and Barr, A. R. (1971). *Nature (Lond.)* **232,** 657–658.

de Zulueta, J., Chang, T. L., Cullen, J. R. and Davidson, G. (1968). *Mosquito News* **28,** 499–503.

SUBJECT INDEX

A

Accessory gland
Normal and atrophied, 59, 75, 76, 77, 92
Role of, in
Monogamy, 19, 20, 49, 75, 92
Oviposition, 19, 20
Aceratagallia sanguinolenta
Susceptibility to potato yellow-dwarf virus, 115
Acraea encedon
Meiotic drive and sex-distortion, 128
Aedes aegypti (yellow fever mosquito)
Accessory gland and monogamy, 19
Chemosterilants
Apholate and inhibition of accessory gland, 59
Laboratory trials, 52, 57, 58, 59
Persistence of metepa in, 51
Resistance, 59
Thiotepa and susceptibility to *P. gallinaceum*, 58
Female-sterile mutant, 129
General genetics of, 118
Genetics of susceptibility to parasites
Filarial worms, 115, 129
P. gallinaceum, 115
Irradiation
Dominant lethals, 16
Field trial, 46
Oogenesis, 16
Resistance, 47–48
Sperm inactivation, 16

Juvenile hormone treatment, 49
Mass-rearing, 10
Meiotic drive and sex-distortion, 128–130
Multiple insemination, 18, 19
Translocations
Double translocation heterozygote, 119
Field release, 119–120
Fitness, 119, 120
Homozygotes, 118, 119
Linkage-group-chromosome correlation, 125
Location of chromosome breakage points, 118
Male-linked, 118
Markers, and the isolation of, 117–118
Multiple, 118
Pseudolinkage, 118
Ae. albopictus
Species competition with *Ae. polynesiensis*, 130–132
Translocations, 122
Ae. hebrideus, 105
Ae. mariae, 97
Ae. pernotatus, 105
Ae. phoeniciae, 97
Ae. polynesiensis
Cytoplasmic incompatibility possibility, 105
Species competition with *Ae. albopictus*, 130–132
Vector of *W. bancrofti*, 130
Ae. pseudoscutellaris, 105

Ae. scutellaris complex
  Hybridization, 105–106
  Member species, 105
Ae. scutellaris katherinensis
  Hybridization with *Ae. s. scutellaris*,
    105–106
Ae. scutellaris scutellaris
  Hybridization with *Ae. s. katheri-
    nensis*, 105–106
Ae. togoi
  Thiotepa and effect on transmission
    of *B. patei* by, 58
Ae. zammitii, 97
Anastrepha ludens (Mexican fruit fly)
  Chemosterilization, 65
  Irradiation, 16, 40
A. suspensa (Caribbean fruit fly)
  Irradiation, 40
Anopheles albimanus
  Cytological recognition of trans-
    locations, 117
  Field trial with chemosterilant
    64–65
  Insecticide resistance, 2
  Mass-rearing, 11
  Population crosses, 97
  Sex-separation of pupae, 11
An. atroparvus
  Hybridization with other *An. macu-
    lipennis* complex species, 70
An. barbirostris, 96
An. claviger, 96
An. farauti No. 1, 96
An. farauti No. 2, 96
An. funestus
  Complex, 96
  Population crosses, 97
An. gambiae complex
  Cage colonisation, 74–75
  Flight activity patterns, 88–89
  Flight sounds, 89
  Identification of member species
    from polytene chromosomes,
    73
  Insecticide resistance, 74
  Malaria transmission, 71–72

Member species, 71–73
Sympatric associations of member
  species, 74
An. gambiae hybrids
  Accessory gland and monogamy,
    19, 20, 75, 92
  Flight activity patterns, 89
  Flight sounds, 89
  Hybrid female contamination, 93–94
  Hybrid female fertility, 75
  Hybrid male competitiveness
    In the field, 80–88
    In the laboratory, 75–79, 89–92
  Hybrid male sterility, 75
  Hybridization
    In the laboratory, 74–75
    In the wild, 74
  Mass-rearing, 12
  Pupal counting, 12
  Sex-distortion, 75, 92–94
  Sex-separation, 13
An. gambiae species A
  Accessory gland and monogamy,
    19, 20
  Distribution, 71–72
  Male reproductive system of hybrid
    with species D, 77
  Mating efficiency selection, 8
  Polytene chromosomes of hybrid
    with species D, 73
  Sex-separation, 13
  Translocations
    Cytological recognition, 117
    Male-linked, three-chromosome
      double, 112
    Y-autosomal, 112
  Upper limit of male mating, 21
An. gambiae species B
  Accessory gland and monogamy,
    19
  Cytological recognition of trans-
    locations, 117
  Distribution, 71–72
  Mass-rearing, 11
  Polytene chromosomes, 72
  Sex-separation, 13
An. gambiae species C

Crosses with species A and B producing sex-distortion, 75, 92
Distribution, 72
An. gambiae species D
Distribution, 72
Male reproductive system of hybrid with species A, 77
Normal male reproductive system, 76
Polytene chromosomes of hybrid with species A, 73
An. hyrcanus, 96
An. koliensis, 96
An. litoralis
Hybridization with *An. sundaicus*, 96
An. leucosphyrus, 96
An. maculipennis complex
Hybrid sterility, 70–71
Malaria transmission, 70, 71
Polytene chromosomes, 71
An. melas
Accessory gland and monogamy, 19, 20
Crosses with species A and B producing sex-distortion, 75, 81, 92
Distribution, 71, 72
Mass-rearing, 11, 12
An. merus
Accessory gland and monogamy, 20
Crosses with species A and B producing sex-distortion, 75, 92
Distribution, 73
An. pharoensis
Population crosses, 97
An. pseudopunctipennis
Population crosses, 97
An. punctulatus complex
Hybrid competitiveness, 96
Hybrid sterility, 96
Member species, 96
An. quadrimaculatus
Accessory gland and monogamy, 19
Comparison of the effects of chemosterilization and irradiation, 63–64
Field trial of irradiation, 45

Mass-rearing, 11
Persistence of metepa in, 51
Population crosses, 97
An. sacharovi
Population crosses, 97
An. stephensi
Accessory gland and monogamy, 19
Hybridization with *An. superpictus* 96
Mass-rearing, 11
Population crosses, 97
Susceptibility to *P. gallinaceum*, 116
An. sundaicus
Crosses with *An. litoralis*, 96
Population crosses, 97
An. superpictus
Crosses with *An. stephensi*, 96
An. umbrosus, 96
Anthonomus grandis (Cotton boll weevil)
Cotton pest, 2
Effect of hempa, 52
Field trial with apholate, 66
Field trial with irradiation, 41
Mass-rearing, 8
Aphis gossypii
Cotton pest, 2
Apis mellifera (Honey bee)
Optimum irradiation dose, 31
Assortative mating, 22, 43, 64, 122
Attractant-chemosterilant (= chemosterilant-bait) combination, 24, 51, 53–55, 68
Attractants
For houseflies, 53, 54, 55
Leucine, 68
Light, 51, 60, 61, 67
Methyl eugenol, 38
Protein hydrolysate, 40
Sex pheromone, 51
Sound, 51

B

Brugia malayi
Periodic and sub-periodic, 115
Susceptibility of *Ae. aegypti*, 115
B. pahangi
Susceptibility of *Ae. aegypti*, 115

Susceptibility of *C. p. fatigans,* 115
B. patei
Thiotepa and effect on transmission
of by *Ae. togoi,* 58

C

Ceratitis capitata (Mediterranean fruit
fly)
Chemosterilization, 65–66
Irradiation trials
Field, 39, 40
Laboratory, 39
Mass-rearing, 8, 9
Polygamy, 39
Chemosterilants
Accessory gland substance, 49
Advantages, 51
Alkylating agents, 49, 50
Antimetabolites, 49
Application methods, 51, 55, 67
Aziridines, 50
Chemical mutagens, 50
Common chemosterilants, e.g., tepa,
metepa, thiotepa, apholate (see
under individual insect species)
Disadvantages, 51
Hempa (= hexamethylphosphor-
amide), 52
Hexamethylmelamine, 52
Insecticides, 49
Juvenile hormones, 49
Methiotepa, 56
Myleran, 51
N, N′-tetramethylenebis (1-aziridine
carboxamide) (= TMAC), 55
Non-alkylating, 52
Organo-metals, 49
Other chemicals, 49
Persistence in treated insects, 51, 52
p, p-bis (1-aziridinyl)-N-methyl
phosphinothioic amide, 64–65
Toxicity hazards, 51
Tretamine (= melamine), 50
Tri(ethyleneimino)triazine
(= TEM), 50, 51
Cicadilina mbila

Susceptibility to maize streak-disease
virus, 115
Clunio
Unidirectional incompatibility, 105
Cochliomyia hominivorax (Screw-worm)
Chemosterilization
Hempa, 52
TMAC, 55
Flight ranges, 33
Irradiation
Comparison with tretamine, 50
Field campaign, 32–36
Male and female releases, 18
Optimum dose, 31, 32
Sexual aggression test, 33
Mass-rearing, 9
Translocation, 124
Compound chromosomes
D. melanogaster, 125
Conditional lethal genes, 114, 116,
127–128, 130
Cotton spraying, 2
Culex pipiens australicus
Distribution, 100
Culex pipiens complex
Filariasis, 100
Hempa, effect of, 52
Member species, 99
Species and population crosses, 100
Translocations
Cytological recognition of, 120
Field release, 121
Homozygous, 120
Integration with cytoplasmic in-
compatibility, 121–122
Isolations, 120
Linkage-group-chromosome cor-
relation, 125
Multiple, 120, 121
C. p. fatigans
Chemosterilants
Autosterilization experiment with
tepa, 60–61
Field trials, 61–63, 104–105
Laboratory trials, 59
Large-cage trials with apholate
and tepa, 60

Persistence of thiotepa and tepa in, 52

Cytoplasmic incompatibility
Integration with translocations, 105
Population elimination, field trial, 102–105
Population replacement, 105
Distribution, 100
Field trial with irradiation, 45, 46, 47
Mass-rearing
Pupal separation, 10
Sex-separation, 10, 11
Reproduction rate, 28, 29
Susceptibility of filarial worms, 105, 115

C. p. molestus
Systematic position, 99

C. p. pipiens
Accessory gland and monogamy, 19
Distribution, 99
Transmission of P. cathemerium, 115

C. tarsalis
Persistence of metepa in, 51–52

C. tritaeniorhynchus
Linkage-group-chromosome correlation, 125
Translocations, 122
Vector of Japanese B encephalitis, 122

Cytoplasmic incompatibility
Ae. scutellaris complex
Distribution, 105
Hybridization, 105–106
C. pipiens complex
Bidirectional incompatibility, 100, 102, 103
Field trials, 102–105
Incompatibility determinants, 101
Integration with translocations, 105, 121–122
Meiotic parthenogenesis, 101
Partial incompatibility, 101
Population replacement mechanism, 105
Unidirectional incompatibility, 100, 102

D

Dacus cucurbitae (Melon fly)
Chemosterilization, 65–66
Irradiation, field trial, 37
D. dorsalis (Oriental fruit fly)
Chemosterilization, 65–66
Irradiation, field trial, 37
Male annihilation, 38
D. oleae (Olive fly)
Chemosterilization
Apholate-bait field trial, 68
Laboratory trials, 65–66
Irradiation, 40
D. tryoni (Queensland fruit fly)
Irradiation, field trial, 40, 41
D. zonatus (Guava fruit fly)
Irradiation, 40
Dermatobia hominis (Torsalo)
Irradiation, 37
Diapause, 98, 116, 127
Diparopsis castanea (Red bollworm)
Chemosterilization
Effect of hempa, 52, 66–67
Persistence of tepa in, 52, 67
Sensitivity compared with housefly, 66
Sterility indices for various chemosterilants, 67
Dirofilaria immitis
Susceptibility of Ae. aegypti, 115
D. repens
Susceptibility of Ae. aegypti, 115
Drosophila (melanogaster)
Alkylating agents, 50
Compound chromosomes, 125
Mating efficiency selection, 8
Meiotic drive, 128
Mutagenic chemicals, 50
Translocation homozygotes, 109

E

Environmental pollution, 2, 133
Erioischia brassicae (Cabbage root fly)
Large-cage tepa-bait trial, 68

F

Fall armyworm

Cotton pest, 2
Farex, 11
Fruit flies (see also under specific names)
Effect of hempa, 52
Female polygamy, 20
Mass-rearing, 9

G

Glossina austeni
Mass-rearing, 13
Female polygamy, 20
Irradiation, 37
Translocations, 123
G. fuscipes
Mass-rearing, 13
G. morsitans
Chemosterilization
Field trial with tepa, 56–57
Laboratory trials, 56
Female polygamy, 20
Hybridization
Between "subspecies", 98
With *G. swynnertoni*, 97
Irradiation, 36, 37
Mass-rearing, 13, 14
Upper limit of male mating, 21
G. swynnertoni
Hybridization with *G. morsitans*, 97
G. tachinioides
Irradiation, 37
Mass-rearing, 13

H

Habrobracon juglandis
Irradiation
Dominant lethal mutations, 14, 15, 16
Dose fractionation, 31
Sperm inactivation, 16
Temperature effect, 32
Nitrogen mustard aerosol, 16
Haematobium irritans (Hornfly)
Irradiation, field trial, 37
Heliothis zea (Corn earworm)
Irradiation, field trial, 44
Hippelates pusio (Eye-gnat)

Optimum irradiation dose, 31
Holokinetic chromosomes, 42
Hybrid sterility
Artificial (= forced) mating technique, 70
Hybrid vigour, 69, 75, 134
Interspecific crosses, 69
Mating barriers, 69, 88, 134
Species definition, 69
Hylemya antiqua (Onion fly)
Apholate-bait, large-cage trial, 68
Irradiation, 41
Mass-rearing, 41
Hypoderma lineatum (Cattle grub)
Irradiation, 37

I

Insect control
Attractant traps, 3, 133
Biological methods
Parasites, 3
Pathogens, 3
Predators, 3
Chemicals, 1
Crop culture practices, 3, 133
Environmental sanitation, 3, 133
Genetic means, 3, 4, 133
Juvenile hormones, 3, 133
Resistant crop plants, 3, 133
Insect releases
Place, 17, 135
Sex, 18, 21, 22, 27, 134, 135
Time, 17, 29–30, 135
Insecticide resistance, 1, 2, 74, 133
Insecticides
Azinophosmethyl, 2
Azodrin, 2
BHC, 2, 56, 57
Carbaryl, 2
DDT, 2, 74
Diazinon, 2
Dichlorvos, 2, 54
Dieldrin, 2, 54, 57, 74
Dimethoate, 2, 53
Fenitrothion, 2
Fenthion, 2
Isolan, 2

Malathion, 2, 40
Methyl parathion, 2
Naled, 2, 38
Parathion, 2
Propoxur, 2
Pyrethrins, 2
Ronnel, 2
Trichlorfon, 2, 54
Zytron, 2
Irradiation (see also under specific insect pests)
   Atmosphere effect, 32, 47
   Fractionation of delivery, 32
   Optimum dose rates, 31, 32
   Sources, 31

L

Laspeyresia pomonella (Codling moth)
   Chemosterilization
      Effect of hempa, 52
      Field trial with tepa, 68
   Female polygamy, 20
   Irradiation
      Field trial, 42, 43
      Optimum dose, 31, 42, 43
   Multiple translocations, 124
Lepidoptera (see also under individual species)
   Irradiation, 41–42, 124
   Multiple translocations, 42, 124
   Sex-determination, 128
Lethal factors, 127
L. cuprina (Sheep blowfly)
   Booby-trapping, 54
   Female monogamy and polygamy, 21
   Linkage-group-chromosome correlation, 125
   Male and female releases, 21
   Optimum irradiation dose, 21
   Sex-separation, 21
   Upper limit of male mating, 21
L. sericata (Blowfly)
   Female polygamy, 45
   Irradiation, field trial, 45
   Translocations, 123
   Upper limit of male mating, 45

Lymantria dispar (Gipsy moth)
   Irradiation, field trial, 44

M

Male accumulation, 30
Male mating, upper limit of, 20, 21, 45
Male mosquito survival, 30
Mass-rearing
   Automation, 7
   Colonies, 6, 134
   Economics, 8
   Feeding, 6
   Hybrid vigour, 7
   Inbreeding, 7
   Insect quality, 7, 134
   Mating, 8
   Outcrossing, 7
   Overcrowding, 8
   Oviposition, 7
   Pathogen control, 6
   Sex-separation, 7, 130
Mating sequences
   Copulation, 18, 99
   Courtship, 18
   Gamete storage, 18, 99
   Insemination, 18, 99
   Ovum fertilisation, 99
Matrone, 19
Meiotic drive
   A. encedon, 128
   Ae. aegypti, 128–130
   D. melanogaster, 128
   Sex-separation aid, 130
   Transport mechanism, 127, 129–130, 134
Melolontha vulgaris (Cockchafer)
   Irradiation, field trials, 41
Mormoniella
   Unidirectional incompatibility, 105
Musca domestica (Housefly)
   Chemosterilization
      Booby-trapping, 54, 55
      Field trials
         Apholate, 53, 54
         Hempa, 53, 54, 55
         Metepa, 53, 54, 55
         Tepa, 53, 54, 55

TMAC, 55
Laboratory trials
  Apholate, 52, 53
  Hempa, 52
  Metepa, 27, 51–52
  Repellency of metepa, 56
  Resistance, 56
  Synergism between tepa and hempa, 56
Insecticide resistance, 2, 3
Irradiation, field trial, 45
Mass-rearing, 12
Monogamy, 20
Reproduction rate, 27
Sex-distortion, 130
Sex-separation, 12, 13, 130
Translocations
  Isolation, 123
  Large-cage trials, 123
  Linkage-group-chromosome correlation, 125
Musca domestica vicina
Hempa, field trail, 55

O
Oncopelta fasciatus (Milkweed bug)
Effect of, hempa, 32

P
Paramyelois transitella (Navel orangeworm)
Female polygamy, 20
Pectinophora gossypiella (Pink bollworm)
Metepa, large-cage trials, 68
Pilot field trials, 135–136
Plasmodium berghei yoelii
Genetic recombination in, 116
P. cathemerium
Transmission by *C. p. pipiens*, 115
P. gallinaceum
Susceptibility of *Ae. aegypti,* 115
Susceptibility of *An. stephensi,* 116
Thiotepa and effect on transmission of by *Ae. aegypti,* 58
Popillia japonica (Japanese beetle)
Effects of hempa, 52
Population dynamics (see also under Translocation dynamics)

Density dependence, 24, 25, 27, 30, 134
Insect population production parameters, 25, 26, 28, 29, 30
Integrated control, 22, 23, 26, 135
Mathematical models and computer simulation, 23–30, 82, 83, 86, 88
Reproduction rate (= rate of population increase = reproductive potential = net reproduction rate), 24, 25, 27, 28, 29
Pseudolinkage, 118
Pyrrhocorris apterus (Lime bug)
Juvenile hormone treatment, 49

R
Radiation resistance, 47–48
Resistance to genetic control, 7, 27
Rhagoletis cerasi (Cherry fruit fly)
Irradiation, 40
Rhodnius neglectus
Hybridization studies, 98
R. prolixus
Hybridization studies, 98
Irradiation
  Atmosphere effect, 47
  Optimum dose, 31, 47
Mass-rearing, 14
Metepa treatment, 65

S
Sex-determination
  Anophelines, 109
  Culex and *Aedes,* 109, 118, 129
  Lepidoptera, 128
  M. domestica, 130
Sex-distortion
  A. encedon, 128
  Ae. aegypti, 128
  An. gambiae hybrids, 75, 92–94
  An. maculipennis hybrids, 70
  D. melanogaster, 128
  M. domestica, 130
Sibling species
  An. gambiae complex, 95
  An. maculipennis complex, 71
Species replacement, 130–132

Spodoptera litoralis (Egyptian cotton leafworm)
  Cotton pest, 2
  Insecticide resistance, 2
  Sterilization by irradiation and chemosterilants
    Aspermia, 16, 21, 50
    Chromosome breakage, 15–16
    Different effects of irradiation and chemosterilants, 50
    Effect in relation to life-history stage, 17
    Effect in relation to sex, 17
    Effect on mating competitiveness, 15, 16, 17, 26, 50
    Effect on oogenesis, 16, 50
    Effect on spermatogenesis, 14, 50
    Effect on spermatozoa competitiveness, 17, 21, 26
    Genetic imbalance, 16
    Production of dominant lethal mutations, 14, 16, 21, 31, 32, 49–50, 133
    Recovery of fertility, 17
    Somatic damage, 14, 50
    Sperm inactivation, 16, 21, 50
Stomoxys calcitrans (Stable fly)
  Irradiation, field trial, 37, 38
  Methiotepa field trial, 56
  Reproduction rate, 28

T

Teleogryllus commodus
  Hybridization with *T. oceanicus*, 98
T. oceanicus
  Hybridization with *T. commodus*, 98
Translocation dynamics
  Density dependence, 114, 122, 123
  Double translocation heterozygote, 114, 119
  Multiple translocations, 113, 114
  Population elimination, 114, 133
  Population replacement, 114, 115, 116, 133
  Simple translocations, 113, 119
Translocations
  Adjacent—1 segregation, 107, 108

Adjacent—2 segregation, 107, 108
Alternate segregation, 107, 108
Aneuploid (unbalanced) gametes, 107
Autosome-autosome translocation, 109, 117
Definition, 107
Isolation, 116–124
Multiple translocation, 112
Orthoploid (balanced) gametes, 107
Semi-sterility, 107, 117
Simple translocation, 108
Translocation heterozygote, 107, 108, 109, 113, 117
Translocation homozygote, 108, 109, 113, 117
Viability of translocation homozygotes, 109
X-autosomal translocation, 109–112, 117
Y-autosomal translocation, 109, 112, 117
Triatoma brasiliensis
  Hybridization studies, 98
T. guasayana
  Tepa treatment, 65
T. infestans
  Hybridization studies, 98
  Tepa treatment, 65
T. maculata
  Hybridization studies, 98
T. patagonica
  Tepa treatment, 65
T. sordida
  Hybridization studies, 98
Tribolium castaneum (Flour beetle)
  Insecticide resistance, 3
Trichoplusia ni (Cabbage looper)
  Chemosterilization
    Autosterilization trial, 67
    Laboratory trials with tepa, 67
  Cotton pest, 2
  Irradiation
    Large-cage trials, 44, 124
    Optimum dose, 43, 44
  Multiple translocations, 124
Trypanosoma congolense

Chemosterilization of *G. morsitans* and transmission of, 56

T. cruzi
Chagas' disease, 47

Tsetse flies (see also under specific names)
Chemosterilants, field trials, 56–57
Irradiation, laboratory trials, 36, 37
Mass-rearing, 13, 14
Reproduction rate, 36

W

WHO/ICMR Research Unit on the Genetic Control of Mosquitoes, 10, 47, 62, 103, 104, 105, 119, 122

Wuchereria bancrofti
Aperiodic and *Ae. polynesiensis*, 130
Periodic and sub-periodic, 115
Susceptibility of *Ae. aegypti*, 115